RADIO TELESCOPES

RADIOTELESKOPY

РАДИОТЕЛЕСКОПЫ

The Lebedev Physics Institute Series

Editor: Academican D. V. Skobel'tsyn

Director, P. N. Lebedev Physics Institute, Academy of Sciences of the USSR

Proceedings (Trudy) of the P. N. Lebedev Physics Institute

Volume 28

RADIO TELESCOPES

Edited by
Academician D. V. Skobel'tsyn
Director, P. N. Lebedev Physics Institute
Academy of Sciences of the USSR, Moscow

Translated from Russian

Springer Science+Business Media, LLC 1966

ISBN 978-1-4899-4778-9   ISBN 978-1-4899-4776-5 (eBook)
DOI 10.1007/978-1-4899-4776-5

The Russian text was published by Nauka Press in Moscow in 1965 for the
Academy of Sciences of the USSR as Volume XXVIII of the Proceedings
(Trudy) of the P. N. Lebedev Physics Institute.

Радиотелескопы

Труды Физического института им. П. Н. Лебедева. Том XXVIII

Library of Congress Catalog Card Number 66-14739

PREFACE

The present volume of the Transactions of the P. N. Lebedev Physics Institute of the Academy of Sciences of the USSR (FIAN) includes an account of the work performed in the Radio Astronomy Laboratory during recent years.

It is a well-known fact that the main factors limiting our knowledge of the universe obtained by radio-astronomical methods are the sensitivity and resolving power of radio telescopes.

A high-precision radio telescope, 22 m in diameter, was built some years ago and has been in constant use ever since. Its principal characteristics and the results of some investigations performed with this telescope were published in Volume 17 of these Transactions, which appeared in 1962.

It is now several years since the design and construction of the largest radio telescope in the world (80,000 m^2 in area) began at the Serpukhovskaya Station of FIAN. This radio telescope is the fruit of the work of many scientists, engineers, and technicians. The properties of this radio telescope are discussed in the first two articles of the present collection.

Research is continuing in the Laboratory on various new large radio telescope systems. Novel constructions are being designed, special electronic and radio-frequency circuits are being developed, and new operating methods for the already existing radio telescope are being perfected. These problems are discussed in articles 3 to 6, 8 to 11, 16, and 17. The design of new equipment is considered in article 12, and articles 7 and 13 to 15 cite the results of certain specific investigations.

In spite of its considerable size, the present collection does not give a full indication of all the scientific research being conducted in the Laboratory. Such fields of study as the investigation of the supercorona of the sun or the meter waveband, the investigation of planets, and others, are not reflected in this collection because of limited space. The interested reader may find them in the periodical literature. However, the present collection does, in a sense, sum up a stage in investigations associated with large telescopes and, in our opinion, will be of interest not only to scientists directly concerned with the field of radio astronomy, but also to those specializing in antenna techniques.

V. V. Vitkevich
Director
Radio Astronomy Laboratory

PUBLISHER'S NOTE

The following journals cited in this book are available in cover-to-cover translations:

Russian Title	English Title	Publisher
Radiotekhnika	Telecommunications and Radio Engineering	Institute of Electrical and Electronics Engineers
Radiotekhnika i élektronika	Radio Engineering and Electronic Physics	Institute of Electrical and Electronics Engineers
Astronomicheskii zhurnal	Soviet Astrononomy	American Institute of Physics
Doklady Akademii Nauk SSSR	Soviet Physics–Doklady	American Institute of Physics
Pribory i tekhnika éksperimenta	Instruments and Experimental Techniques	Instrument Society of America
Uspekhi fizicheskikh nauk	Soviet Physics–Uspekhi	American Institute of Physics

CONTENTS

vii

DESIGN PRINCIPLES OF THE FIAN CROSS-TYPE
WIDE-RANGE TELESCOPE

V. V. Vitkevich and P. D. Kalachev

The construction of the largest radio telescope in the world* (with a cross-type antenna system covering a geometric area of 80,000 m²) (Fig. 1) is nearing completion at the P. N. Lebedev Physics Institute of the Academy of Sciences of the USSR (FIAN). The radio telescope is intended for meter waveband operation, but will later be used also in the decimeter waveband. The East—West antenna array is a 40 × 1000 m parabolic cylinder whose elevation can be controlled. The second parabolic cylinder, in the North—South direction, has identical dimensions. Its declination polar-pattern will be electronically controlled. Let us now consider some of the scientific problems for which the above radio telescope will be used.

Scientific Problems

The observational capabilities of modern radio astronomy are, as is well known, determined principally by the sensitivity and resolving power of available radio telescopes. In the meter waveband, the most acute problem is the resolving power, so that discontinuous systems must be used. An example of such a system is the "Mills Cross" whose resolving power is determined by its linear dimensions (about 10' at 3 m, in our case), its effective area being approximately equal to its geometric area.

Such an instrument was selected for the main radio telescope of the Serpukhovskaya radio astronomy station of the FIAN. This design combines high resolving power with wide range, makes possible subsequent operation on shorter waves, and is flexible in operation: it is suitable for forming radiointerferometer bases of various orientations and sizes. The latter capability is achieved by proper selection of the basic system; the energy collected by the individual antenna arrays into which the East—West and North—South lines are divided by radio-frequency cables of equal electrical length, is brought into a laboratory located at the center of the radio telescope, where various switching connections can be made. The first basic version of the radio telescope has been designed to operate in the 2.5- to 10-m waveband. The radio telescope is intended for many scientific studies such as the investigation of the near-sun space, the study of radio-frequency radiation from the sun, planets, and other discrete sources. Let us now mention some of the proposed investigations.

The telescope described will be capable of detecting thousands of discrete sources and will also create new possibilities for the study of the sun's supercorona. Undoubtedly, among the many sources distributed isotropically over the sky, there will be some which are located in the vicinity of the ecliptic. Observation of their radio-frequency radiation during periods of their excitation by the supercorona will make it possible to "transilluminate" the supercorona at various regions and at different times of the year, and thus to obtain much more information about its structure than now available, particularly about the structure of its inhomogeneities. Observations made with interferometer bases oriented in various directions will permit a detailed study of the magnetic field of the sun's supercorona. At first, only the Crab nebula was used to "transilluminate" the supercorona. Recently, however, Slee (Australia) has observed other sources and confirmed the validity of such investigations.

---

*Scientific director of the work — V. V. Vitkevich; chief design engineer — P. D. Kalachev.

Fig. 1. General view of the FIAN radio telescope.

The new radio telescope will make it possible to solve many problems concerning the sun. Radio images of the sun obtained with polarized as well as with unpolarized waves will undoubtedly provide much new information about the structure and dynamics of the corona. In this connection, the use of small, portable, auxiliary antennas will make possible a considerable improvement in the resolving power of the instrument.

The most important task of modern radio astronomy is to investigate the nature and the principal characteristics of discrete radio sources. About two thousand extragalactic sources have been discovered thus far; of this number only about twenty have been identified. To investigate their nature, their spectra, as well as their angular dimensions, must be known.

With the help of the described radio telescope operating at the meter waveband, it will be possible to gather much new information about thousands of discrete sources of extragalactic origin.

An important deficiency of measurements conducted at the present time is the fact that intensity spectra can usually be obtained only as a function of wavelength. However, the lack of information about the angular dimensions of an investigated source makes it impossible to proceed further and, in particular, to find the brightness temperature of the source at the given wavelength, which is a true characteristic of the source. Such information obtained for many thousands of radio sources will contribute significantly to the solution of the problem. The work in this direction which has been done at several places proves its real value.

The solution of such problems will be helped by the use of small portable antennas which are now being designed, and which, together with the main cross type radio telescope, will make it feasible to measure angular dimensions of many thousands of discrete radio sources. Systematic measurements of this kind will be of great value to modern cosmology. Our radio telescope will make it possible to penetrate into the universe far beyond the reaches of the visible, into regions which move with red-shift velocities close to the speed of light and, possibly, to discover the "boundaries" of the universe.

As a result of statistical analysis of radiation sources and their dimensions, it will be possible to obtain not only the number versus intensity function, but also the function number versus effective temperature relationship.

To utilize most efficiently the large area of the telescope for the study of the greatest possible number of radio sources, the effective collecting area of the antenna system must be appropriately divided. The dimensions which have been selected were dictated by an optimum compromise between sensitivity and resolving power. This has been analyzed by one of the authors [1,2], and (later) developed by Horner [3]. Using the results of the above-mentioned works, it is possible to find as a function of wavelength the number of discrete sources per steradian which can be detected on the basis of the available sensitivity and resolving power. The results of such calculations made by Dagkesamanskii are shown in Fig. 2.

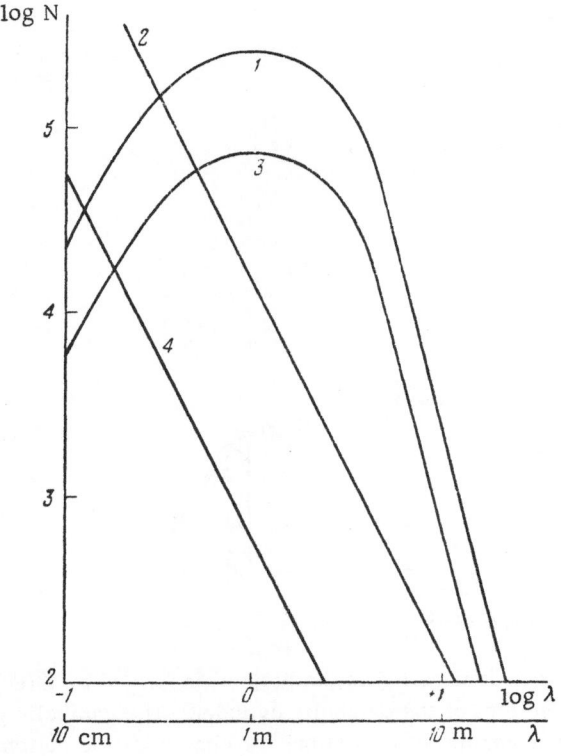

Fig. 2. Number N of sources observable in one steradian as a function of wavelength λ.

The results were calculated for the system operating both as a cross-type radio telescope (curves 1,2) and as a single antenna-array telescope (curves 3,4). Curves 1 and 3 have been calculated in terms of sensitivity, and curves 2 and 4 are in terms of resolving power.

The following basic data have been assumed in the calculations.

a. The sensitivity has been computed under the assumption that no amplifying probes are used with the antenna arrays, and that, because of damping, the effective area is 15% × 0.6 = 9% of the geometric area.

b. The galactic noise is predominant at longer wavelengths. The noise of the equipment and galactic noise have been assumed according to [3], i.e., $13 \cdot 10^3$, 900, and 400 at 10, 3, and 1 m wavelengths, respectively.

c. For reference, the number of sources has been assumed as being 15 per 1 steradian with an energy $2 \cdot 10^{-25}$ (λ = 1.9 m). The rise of intensity with wavelength has an exponent of 0.8 and its magnitude is equal to 1.8 (exponent of the number-to-flux density ratio).

d. The number of sources which can be detected in terms of resolving power is assumed to be such that one source falls within an angle 20 times greater than the solid angle of the antenna pattern at the given wavelength.

It may be seen from the curves shown that when operating with the cross antenna, the agreement between the curves representing the resolution-limited and sensitivity-limited number of sources is quite good.

The first stage of investigations foresees an attempt at polarization studies. For this purpose, multichannel instrumentation operating at adjacent frequencies is being designed. * Because of the different rotation of the polarization plane in the ionosphere at different wavelengths, the relative radiation intensities of polarized sources should vary uniformly as a function of wavelength. Such investigations may prove very interesting if positive results are obtained.

In spite of the various radio investigations of the galactic structure, many essential problems remain still unsolved. Thus, according to Mills (report at the Paris Symposium in 1959), much of the radiation originating in the galactic arms takes place not only at right angles to their axes, but also approximately in parallel to them, which contradicts the concept of magnetic-retardation radiation. Verification and improvement of the accuracy of these results is very important. Such investigations can be performed very efficiently with our telescope.

The described radio telescope will also be used for studying radio emission from the planets, for example, the irregular radiation of Jupiter at meter wavelengths.

Let us now discuss briefly the construction of the separate units of the radio telescope.

_____

*Under the direction of V. A. Udal'tsov.

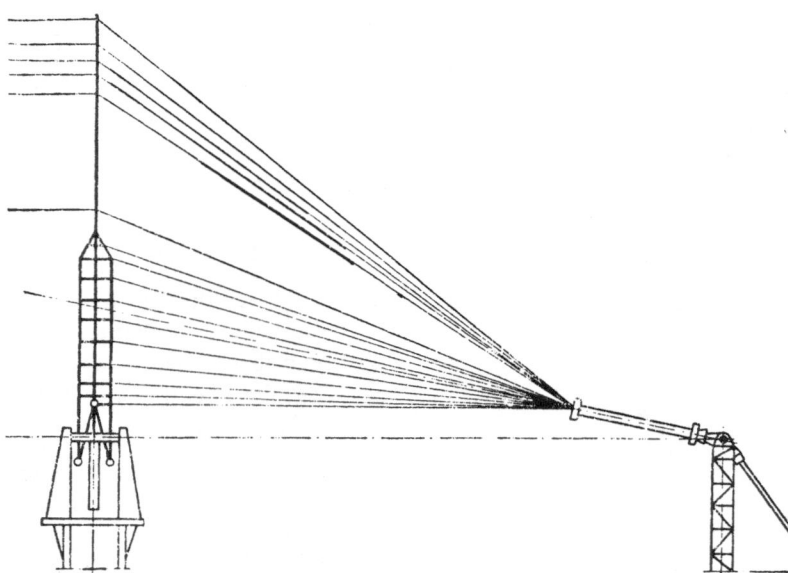

Fig. 3. Stretching arrangement.

The steerable parabolic cylinder, or the steerable line as it is called in the project, is formed by three-dimensional parabolic girders located in the normal cross sections of the parabolic cylinder. The parabolic girders, constructed of welded thin-walled steel tubes, have a triangular cross section: two booms at the bottom and one at the top. Special bosses are welded to the top boom for holding the wires which pass along the cylinder axis in place. The wires are spaced 100 mm apart, and under a definite tension form the reflecting plane of the parabolic cylinder. The wires are fixed in the lateral direction (for keeping the spacing constant) and are free to slip lengthwise in their sockets. They are fastened to both end girders by means of wire-stretching bolts with left- and right-handed threads which control the tension of the wires.

The overall number of wires is 431. The total pull of the wires, amounting to about 50 tons, is absorbed by two anchor supports located behind the two end girders. A system of steel cables is stretched between the end girder and anchor support. The cables are fixed to the top boom of the parabolic girder at intervals of about 2 m. There are 21 such steel cables; they converge in a fanlike manner at one point and are connected to a steel lug. Eight 32-mm steel cables, in the form of a 5-m long braid, pass from the steel lug to the anchor support. The anchor end of the braid terminates in a second steel lug which is fixed to the anchor support by means of a vertical lug and a horizontal swing bolt.

The swing bolt assists in the free sag of the steel braid and of the entire stretching system. When the elevation of the parabolic girders is varied, the steel braid twists and the torsion of the cable braid is absorbed by the separate steel cables as an additional pull, which at a rotation of $\pm 60°$ amounts to about 10% of the main tension.

The steerable parabolic cylinder is rotated by means of an electromechanical drive mechanism provided at each girder. All in all, there are 37 such parabolic girders; the distance between two adjacent girders is 28 m. This distance was dictated by the maximum permissible sag of the wires which form the parabolic cylinder. These considerations also influenced the choice of the high-strength steel wires 2 mm in diameter. At a tension in the wire $N = 50$ kg$_f$, the sag amounts to 50 mm.

A segment gear with a pitch-circle radius $R_{p.c.} = 2448$ mm is provided at the center of each parabolic girder and is engaged with the output pinion of a drive reducer mounted on the top platform of the support pillar. The reducer has an output shaft for mounting a handle of a manual drive mechanism which is used during construction and repair work, and also for initial adjustment of the parabolic girder.

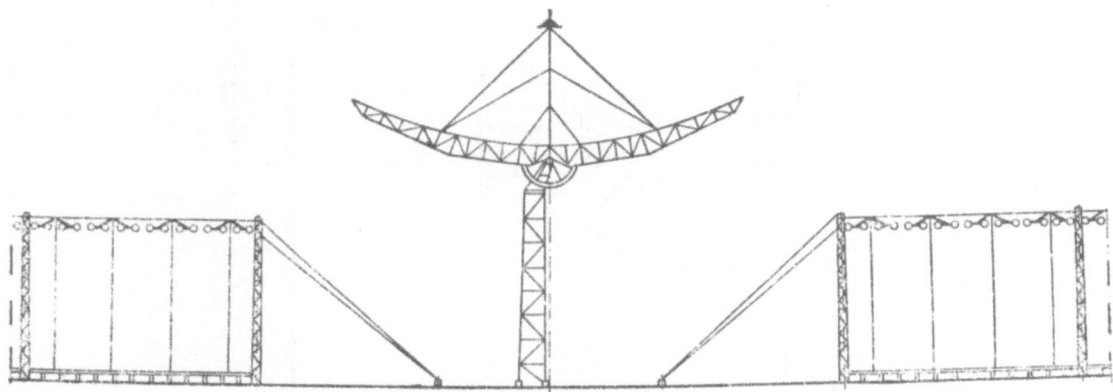

Fig. 4. Intersection of the steerable and fixed lines.

Synchronous rotation of all 37 girders is provided by an electrical drive shaft. The mismatch between the angles of rotation of any two parabolic girders is within 4', corresponding to a linear misalignment of 30 mm at the girder ends.

The instruments for monitoring the rotation of the parabolic girders (74 pcs — two instruments per girder, for coarse and fine indication of the angle of rotation) are located on a control-console panel inside the laboratory. These instruments indicate the position of the parabolic girders at any moment. Selsyn motors coupled to the rotating shafts of the parabolic girders serve as transducers.

If the synchronous rotation of all parabolic girders is disturbed and the angle of mismatch of the relative position of the girders exceeds allowable limits, the system automatically disconnects the supply of electrical power. If the electrical drive shaft does not operate correctly, the parabolic girders may be rotated into any position by means of pushbuttons which control the drive motors of the individual girders, and which are located on the central control panel, the position of the girders being read from the monitoring instruments.

The fixed line of the cross antenna is located in the North—South direction and intersects with the steerable line at right angles. Since the steerable line is continuous, the fixed line is broken at the intersection (Fig. 4).

The reflecting parabolic cylinder of the fixed line is formed by crosswise wires spaced every 100 mm. A railway in the form of two I-beams spaced 1.8 m apart passes along the center of the parabolic cylinder. The ends of the crosswise wires which form the parabolic cylinder are fixed to lengthwise thin-walled tubes suspended from the support pillars of the lengthwise end rows.

One end pillar is shown in Fig. 5. Here, 1 is the pillar, 2, 3, 4, and 5 are the pillar braces, 6 is the lengthwise tube, and 7 are the crosswise wires. The pillars are located 10 m apart.

The spacing between the crosswires is kept constant by six lengthwise 10-mm rods. The crosswise wires pass through 2.2-mm holes drilled in the rods every 100 mm. The rods are fixed to the intermediate support pillars spaced, as are the end pillars, every 10 m. The lengthwise rods are fixed to the intermediate pillars so that their height above ground may be varied.

The dipole feeder system is suspended from the dipole masts. The dipole masts of the steerable line, made in the form of welded-pipe trusses, are located at the center of the parabolic girders and are strengthened by three sets of bracing cables. The dipole masts of the fixed line consist of steel trusses formed of welded angle-brackets and are spaced 24 m apart along the railway. The entire cable system is placed into asbestos—cement tubes laid in 1.8-m deep trenches.

The reinforced-concrete beds of the pillars which bear the parabolic girders of the steerable line are hollow and serve as switching booths for the entire cable system.

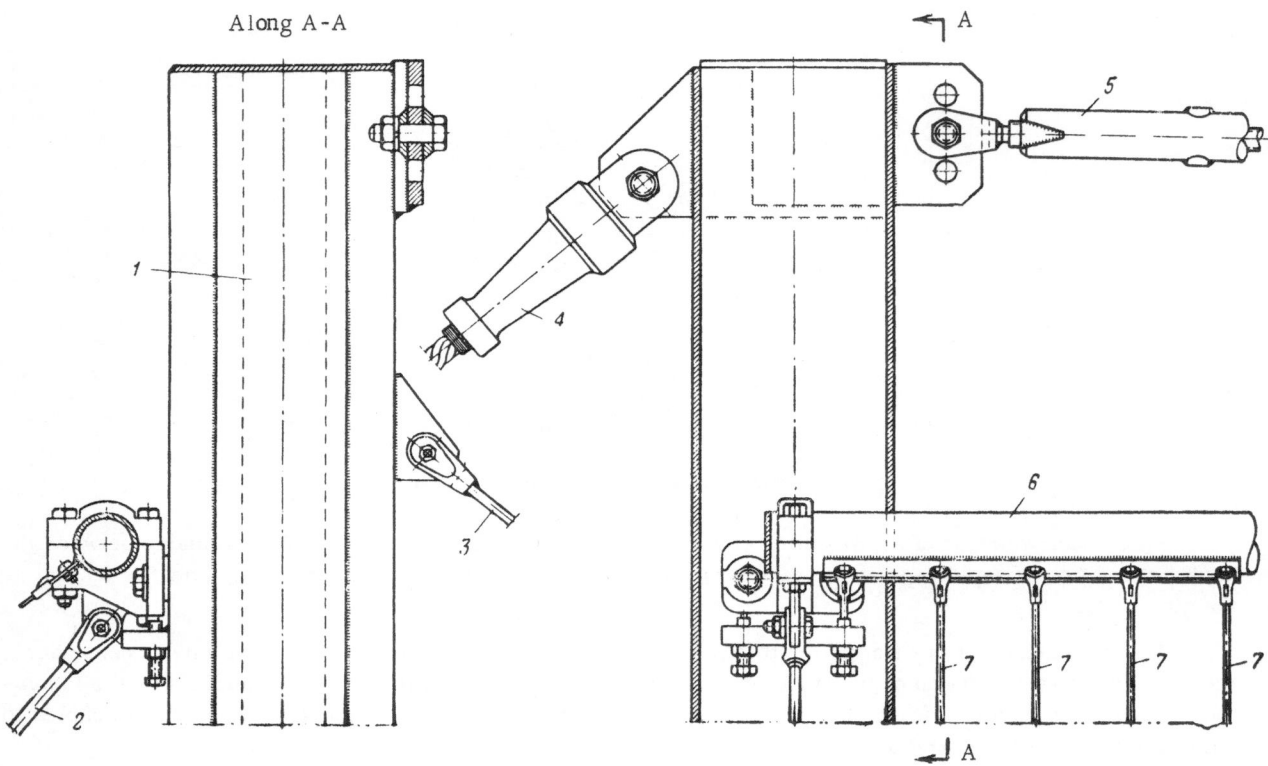

Fig. 5. Corner pillar of the fixed line.

The construction of the steerable East—West line has now been completed. Work is at present being done on the adjustment and test of the control instruments and of the dipole feeding system. The construction and adjustment of the reflector and counter-reflector of the fixed North—South line will be completed shortly.

Literature Cited

1.   V. V. Vitkevich, Astr. Zh., 34(3) (1957).
2.   V. V. Vitkevich, Proceedings of the Sixth Conference on Cosmogony, Izd. Akad. Nauk SSSR (1959).
3.   S. Hoerner, Publ. NRAO, Green Bank, West Virginia, 1(2) (1961).

WIDE-RANGE EXCITER FOR A
"PARABOLIC CYLINDER" ANTENNA

Yu. P. Ilyasov and A. D. Kuz'min

Introduction

As was reported before [1], a large cross-type radio telescope, similar to that described by Mills [2], is being constructed at the P. N. Lebedev Physics Institute of the Academy of Sciences of the USSR (FIAN). The radio telescope consists of two mutually perpendicular antennas of the "parabolic cylinder" type with a focal length of 14.5 m and has the following dimensions: along the generatrix, 1008 m and along the subtending chord, 40 m. The calculated half-power width of the radiation pattern is 13' at a 3.7-m wavelength.

The radio telescope is a meridional instrument. One of its component parabolic cylinders whose axis is oriented in the North—South direction is fixed. The other cylinder oriented in the East—West direction has a variable elevation.

Because of their large size, such antennas are complex and expensive structures. It is thus reasonable to make them capable of operating over a wide range of frequencies. As is well known, a parabolic reflector is a wide-range device and, thus, the problem reduces to designing an appropriate exciter that will also be capable of operating over a wide range of frequencies. This problem has two solutions: it is possible to use either a combined system of several exciters, each one operating at a narrow frequency band, or a single exciter covering the entire frequency range from 30 to 120 Mc.

It is obvious that the use of a combined exciter complicates the antenna design and, furthermore, restricts its spectral-measurement capabilities. More convenient, from the point of view of operability, is a single wide-range exciter. It was thus decided to proceed with the latter exciter system.

It is quite obvious that the exciter of a parabolic cylinder reflector should be located in the focal line and be in the form of a multielement cophased linear system. Cophase excitation of the individual radiating elements in the chosen system is provided by a feeder system in which the electrical length from the receiver input to each element is the same. This is achieved by consecutive splitting of the feeder system together with changing of the characteristic impedance of each new section by a factor of two (Fig. 1). This system will subsequently be called a binary system, since the number of cophase excited elements can only be equal to 2^n, n being an integer. The basic radiator element is a wide-range shunted dipole [3].

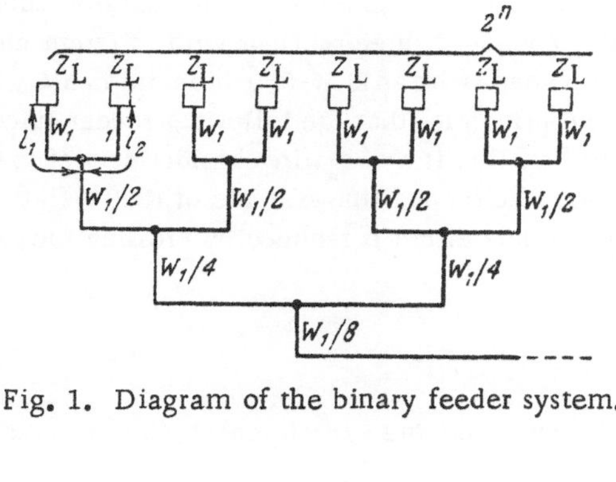

From design considerations, the entire exciter has been divided into 32 identical sections, eight dipoles in each section. The results of the design and experimental investigation of one such section are summarized in the present report.

Fig. 1. Diagram of the binary feeder system.

Radiator Element

Before proceeding with the determination of the shunted dipole parameters, we had to determine the height of the dipole suspension above the counter-reflector. As is well known, the lower the suspension of a dipole, the less its radiation re-

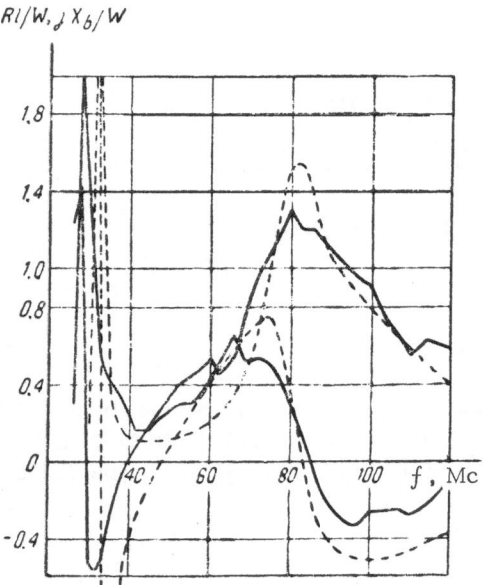

Fig. 2. Feed impedance of a shunted dipole. Height of suspension above the counter-reflector H = 90 cm, W = 600 Ω. At 100 Mc, the two top curves are for R_l/W, and the two bottom ones for jX_b/W.

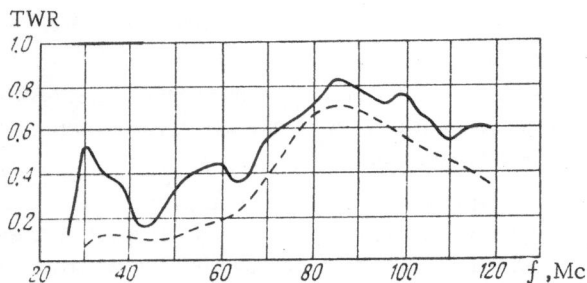

Fig. 3. Traveling-wire ratio (TWR) of a single shunted dipole along a two-wire feeder with W = 600 Ω. Suspension height of dipoles above counter-reflector H = 90 cm.

sistance. On the other hand, the suspension cannot be lower than 0.3 to 0.4 λ, since the radiation pattern will then exhibit a discontinuity in the main direction. To determine the maximum possible suspension height exactly, we have conducted measurements with a 10-cm-wave model of the dipole. It has been found that the first indication of a discontinuity in the radiation pattern appears at a suspension height of 0.36 λ. Starting with this figure, which has been assumed as maximum, and with the shortest wave of the operating interval λ_s = 2.5 m, we have arrived at a suspension height over the counter-reflector of 90 cm. For this suspension height, and taking into account the correction for phase velocity made necessary by the large diameter of the dipole, we have calculated the feed impedance of the dipole as a function of frequency for several types of dipoles with different parameters. The impedance coupled from adjacent dipoles has been neglected. The most suitable for the 30- to 120-Mc frequency band proved a dipole with the following geometric dimensions:

Length of one arm, l_d . . . . 1.5 m
Diameter of dipole, d. . . . . 0.3 m
Number of wires 8
Diameter of wires . . . . 2 mm
Length of shunt. . . . l_{sh} = 0.5 l_d = 0.75 m
Number of wires in shunt. 2

The results of calculating the dipole parameters are shown in Fig. 2.* The experimental and calculated frequency dependence of the traveling-wave ratio (TWR) along a two-wire feeder with a characteristic impedance of 600 Ω loaded by a shunted dipole is shown in Fig. 3. The calculated results agree quite well with the experimental ones and indicate the feasibility of obtaining a TWR greater than or equal to 0.2 in the specified frequency range.

Feeder System

A binary feeder system for eight radiators should, obviously, consist of three sections with different characteristic impedances. Since the connection to one exciter section is made with an RKK-5/18 coaxial cable with a characteristic impedance of 75 Ω, it follows from the system design principles that the following feeder lines should have characteristic impedances of 150, 300, and 600 Ω. The first line is a two-wire shielded feeder made of two RK-3 coax cables whose outside braids were soldered together. The second line is made of KATV flat cable. A two-wire air feeder serves as the third line section. The antenna effect is reduced by crossing the two-wire feeder lines.

---

*In this and following figures, the calculated curves are shown in broken lines, the experimental ones are in solid lines — Editor's Note.

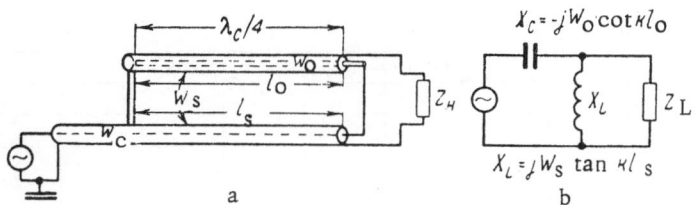

Fig. 4. Circuit (a) and equivalent (b) diagrams of the wide-band
balun (WBB).

Special attention has been paid to the reduction of all possible discontinuities in the feeder system. As was shown in practice, even a small discontinuity along a given path is liable to impair considerably the parameters of a wide-range system. The allowable tolerances in the binary feeder system have been estimated.* It has been found that with a 10% tolerance for the variation of the nominal reflection coefficient, which is between $0 \le |\rho_n| \le 0.67$ ($1 \ge$ TWR ≥ 0.2), the phase mismatch of two identical loads should not exceed $\pm 10°$, and the deviation of the actual characteristic impedance of the feeder should be within $\pm 10\%$ of nominal. If the tolerance of the binary feeder parameters is estimated in terms of the allowable nonuniformities of the amplitude−phase distribution along the exciter dipoles, then, if the allowable nonuniformity of amplitude distribution is assumed to be $\pm 20\%$ and the maximum phase mismatch of the dipole currents is $\pm 15°$, the maximum tolerable difference in the lengths of the binary system dipoles is $\pm 0.0083 \lambda$ ($\pm 3°$). It is this latter tolerance which determines the accuracy of the binary feeder system tuning.

Wide-Band Balun

For coupling the balanced two-wire shielded feeder made of two RK-3 coax cable sections to the RKK-5/18 coax cable we have developed a wide-band balun. The balun is based on the principle of compensating within the desired frequency range the inductive impedance of a quarter-wave two-wire balancing joint by means of a capacitive stub [4-6]. The circuit and equivalent diagrams of such a balun are shown in Fig. 4. Calculations have shown that for the widest band the following conditions must be fulfilled:

$$W_C^2 = W_s W_o \text{ and } W_s \gg W_o,$$

where W_C is the characteristic impedance of the coaxial line, W_s is the characteristic impedance of the short-circuited two-wire stub, and W_o is the characteristic impedance of the open-circuited coaxial stub. The transformation coefficient of the balun is at the same time equal to unity.

The wide-band balun (WBB) has been tested on a decimeter-wave model with a center frequency $f_c = 500$ Mc. The best results were obtained with a WBB having the following parameters: $W_s = 200 \; \Omega$, $W_o = 28 \; \Omega$, and $W_C = 75 \; \Omega$. The input impedance and the TWR of a balun loaded by four terminating absorbers of $75 \; \Omega$ each were measured. The four load resistors were connected to the WBB by means of two sections of a balancing two-wire shielded feeder made of RK-3 cable. The results of these measurements are presented in Fig. 5. After the most important parameters have been determined with the help of the model, a balun was constructed and tested for operation on the meter waveband with a center frequency $f_c = 100$ Mc. The measured data of this balun are shown in Fig. 6. As may be seen in Figs. 5 and 6, the agreement between the calculated and experimental data is quite satisfactory.

Results of Experimental Investigation of One Eight-Dipole Exciter Section

An exciter section consisting of eight dipoles has been built and assembled for field investigations. The height of dipole suspension above metallized ground was 90 cm. The feeder system was pretuned. The tuning consisted in careful adjustment of equal electrical lengths from the dipole ends of the 600 Ω feeder to the balun.

---

*The calculations were made with the participation of chief laboratory technician V. V. Balinov.

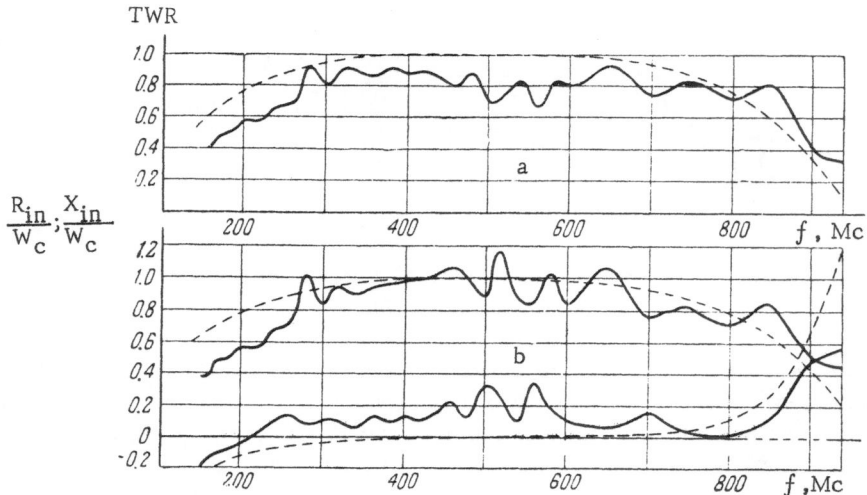

Fig. 5. Experimental and calculated characteristics of the decimeter model of WBB loaded by $W_c = 75\ \Omega$: TWR curves (a), and input impedance curves (b) (normalized to $W_c = 75\ \Omega$).

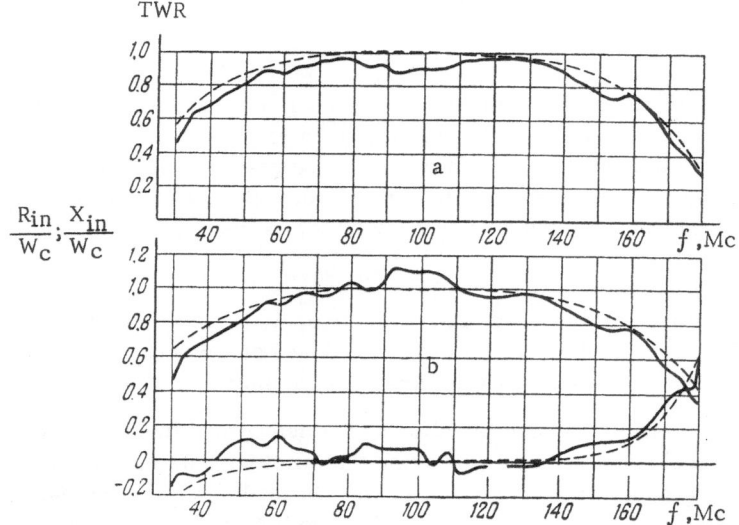

Fig. 6. Experimental and calculated characteristics of the meter waveband WBB (loaded by $W_c = 75\ \Omega$). Notation the same as in Fig. 5.

The phase difference of currents in the absorbing loads at the feeder ends was measured with a single-channel phase meter with a calibrated phase shifter. The loads were phased to within $\pm 4°$ over the entire 30- to 120-Mc frequency range. Special measures were also undertaken to reduce discontinuities in the feeder line. After the feeder system had been carefully tuned, the dipoles were connected to it and the parameters of the entire eight-dipole exciter were measured. The TWR was measured within the frequency range with a measuring 75-Ω coaxial line as well as with an IPSK-2M impedance meter. The TWR of the measuring line proper was determined by the method of measuring small discontinuities and amounted to 0.92 at f = 100 Mc. The length of the measuring line along its slot was 4 m.

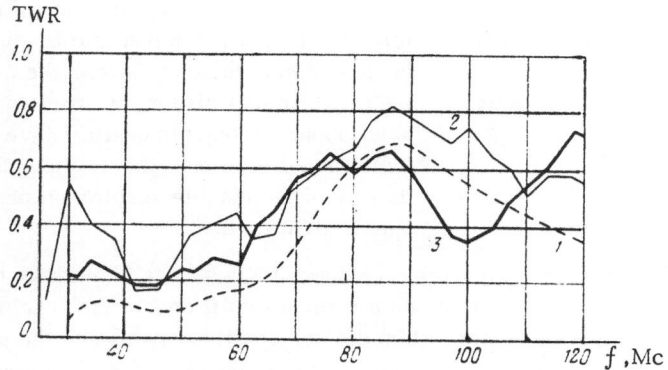

Fig. 7. Frequency dependence of the TWR of an eight-dipole exciter section: calculated (1) , for one dipole (600 Ω feeder) (2), and for eight dipoles with tunable balun (3).

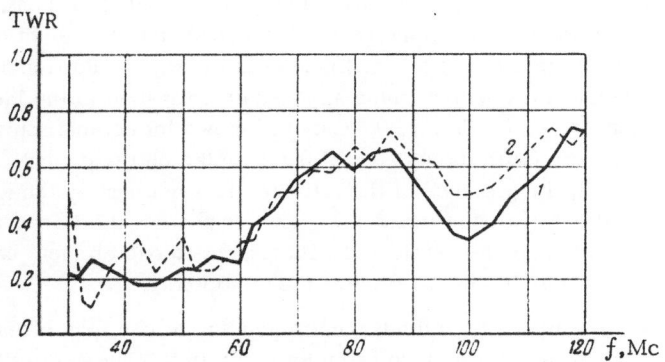

Fig. 8. Measured TWR of an eight-dipole exciter section.

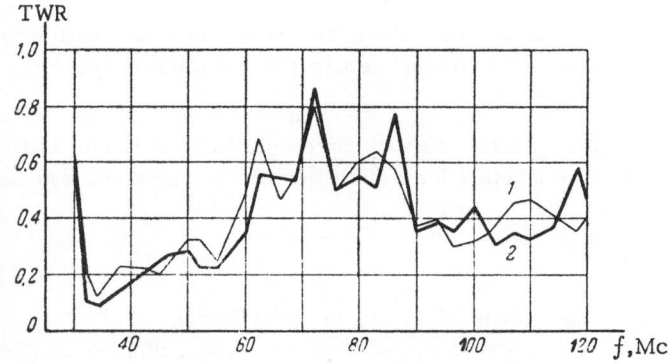

Fig. 9. Measured TWR of an eight-dipole exciter section. 2 × RK-1 cable used in the first feeder line.

f, Mc	S_{geom}, m²	S_{ef}, m²	AUF
47.5	1120	530	0.47
62	1120	705	0.63
84	1120	725	0.65

When determining the TWR of the exciter section, a calculated correction was taken into account for attenuation in the cable coupling the exciter with the measuring line or with the IPSK-2M, so that all the results cited below refer directly to the balun input. A tunable quarter-wave variable-length balun was sometimes used in addition to the wide-band balun. The uniformity of the tunable coupler has been tested by the short- and open-circuit methods.

The TWR of an eight-dipole exciter section is shown in Fig. 7 as a function of frequency. The satisfactory qualitative agreement between the calculated and measured results enable the conclusion to be drawn that a carefully assembled and tuned binary feeder system did not significantly change the performance of the system (curves 2 and 3). Figure 8 shows the results (curve 2) of measuring the TWR of an eight-dipole exciter section coupled to the measuring line by the wide-band balun. Curve 1 has been repeated for comparison with curve 3 of Fig. 7. Comparing the curves shown in Fig. 8, one notes that the use of the balun introduced significant changes only into the long-wave portion of the frequency band.

All the results cited here were obtained with the exciter assembled on the ground. Figure 9 shows the experimental curve 2 which has been obtained with an exciter located at the focus of the reflector. A comparison of curve 2 in Fig. 9 with curve 2 in Fig. 8 shows clearly the difference between them. It seems to us that this is due not to the influence of the reflector but to the fact that for reducing the weight of the feeder an RK-1 cable was used instead of the RK-3 cable in the first feeder line. In this case, an additional small discontinuity appears at the point where the balun is connected, since an RK-3 cable is used in the latter. This is apparently the reason for the large fluctuations in curve 2 in Fig. 9. Curve 1, shown for comparison in Fig. 9, has been obtained with the same exciter section (with the RK-1 cable) assembled on the ground. As may be easily seen from a comparison of both these curves, the presence of the reflector does not destroy the wide-band properties of the exciter section. It may be easily verified that since the system is phased, the effect of the reflector in such exciters with parallel power summing need be considered with respect to one element only, and the TWR of the entire system will be determined mainly by the parameters of one dipole.

An experimental model of a parabolic cylinder, corresponding to one span of the actual antenna, has been built for investigating the problems associated with reflector excitation. The exciter has been mounted in the focal line of this parabolic cylinder, so that the focal line passed between the dipoles and counter-reflector. The principal maximum of the radiation pattern has been trained on the discrete radio source Cassiopeia-A (at its upper culmination). The gain and radiation pattern of the antenna were measured with respect to this source. The results of these measurements, made on three frequencies, are given in the table.

It may be expected from the experimental results obtained that the construction and utilization of the cross-type radio telescope is quite realizable and demands only careful construction and adjustment as well as utilization.

The authors consider it their pleasant duty to express gratitude to Doctor of Engineering Sciences G. Z. Aizenberg and to Engineer V. D. Kuznetsov for a discussion of some problems associated with the design of the system described.

Literature Cited

1. V. V. Vitkevich, Vestn. Akad. Nauk SSSR, No. 5: 23-32 (1961).
2. B. Y. Mills, A. G. Little, K. V. Sheridan, and O. B. Slee, Proc. IRE, No. 1: 67-84 (1958).
3. V. D. Kuznetsov, Radiotekhnika, No. 10 (1958).
4. N. Marchand, Electronics, 17: 142-145 (Dec. 1944).
5. J. W. McLaughlin, D. A. Dunn, and R. W. Grow, IRE Trans., MTT, 5(3): 314-316 (1958).
6. W. K. Roberts, Proc. IRE, 45(12): 1628-1631 (1957).

WIDE-RANGE EXCITER WITH AN ELECTRICALLY SCANNED
RADIATION PATTERN

S. N. Ivanov, Yu. P. Ilyasov, and G. N. Khramov

In this article, we present the results of designing the exciter of the North—South line of the FIAN radio telescope. It is shown that it is possible in principle to design a feeder system in which the systematic errors in amplitude—phase distribution, taking place along the exciter when the dipoles are mismatched with the feeder line in a given sector of the radiation pattern, are reduced considerably. The results of investigating the wide-band properties of the feeder elements, hybrid rings, and of a balancing coupler with a 4:1 transformation ratio, are also cited. Initial experimental results obtained with one exciter section consisting of eight wide-band dipoles are reported.

Introduction

As was already reported [1], the cross-type radio telescope of the FIAN consists of two parabolic cylinders with a generatrix and subtending chord of 1 km and 40 m, respectively. The North—South line of this radio telescope is fixed, and to change the declination of its radiation pattern the phase distribution of the currents flowing in the exciter elements must be controlled, i.e., the antenna beam must be electrically scanned.

The exciter of the East—West line of the KR-1000 cross-type radio telescope is designed for operating within a wide range of frequencies spreading from 30 to 120 Mc. It would thus be very desirable if the exciter system of the North—South line could perform satisfactorily within this frequency range or, at least, within a considerable portion of it. This would make it possible to widen the scope of scientific problems whose solution is sought with the KR-1000 radio telescope, and thus justify the inevitable design difficulties associated with wide-band operation of the exciter. One self-evident requirement is the necessity to keep the level of side lobe radiation as low as possible within the entire scanning sector from 35° north to 85° south of zenith. The accuracy of aiming the radiation pattern should be better than 0.1 of its half-power width in any position.

The exciter of the North—South antenna consists of a line of radiators placed at equal intervals along the focal line of a parabolic cylinder and forms a one-dimensional linear phased array. General properties of such arrays have been fully discussed in [3]. It has been shown that as the angle of inclination of the principal maximum increases, the width of the radiation pattern increases according to

$$2\theta_{0.5C} = \frac{2\theta_{0.5}}{\cos \theta_C},$$ (1)

where θ_C is the angle of inclination reckoned from zenith, and $2\theta_{0.5}$ is the half-power width of the radiation pattern at $\theta_C = 0$. From [3] it follows that the effective area of the antenna will decrease in proportion to $\cos \theta_C$.

In addition to broadening of the radiation pattern, the increase of the angle of inclination results in a growing eccentricity of the radiation beam, i.e., in nonuniform broadening of the radiation pattern on both sides of the principal maximum. The magnitude of eccentricity may be estimated from

$$e = \frac{2\sigma_{0.5}}{8 \cos \theta_C} \tan \theta_C.$$ (2)

Table 1 shows data which characterize the change of antenna parameters with the angle of inclination of the radiation pattern at discrete points of the 2.5- to 10-m waveband in a scanning sector from 0 to 85°.

13

Table 1

θ_C	λ, m	$2\theta_{0.5C}$	e	$k = \dfrac{A_{eff_{\theta_C}}}{A_{eff}}$
0°	3	0°11′	0	1
	6	21.6	0	1
	10	36	0	1
30	2.5	10.5	0′.76	0.866
	3	12.7	0.92	—
	6	25.0	1.8	—
	10	41.5	3	—
60	2.5	18.0	3.9	0.5
	3	22.0	4.8	0.5
	6	43.2	9.3	0.5
	10	1°12′	15.6	0.5
85	2.5	1°44	15	0.087
	3	2 06	18	0.087
	6	4 08	35.5	0.087
	10	6 52	59	0.087

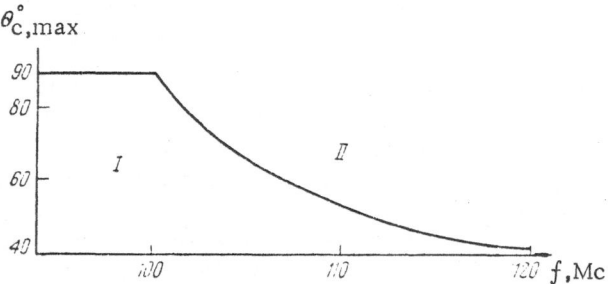

Fig. 1. Maximum angle of inclination of the antenna radiation pattern as a function of frequency. Spacing between elements of the array d = 1.5 m. I, region where no diffraction peaks occur; II, region of diffraction peaks.

The data in the table indicate that the general properties of the antenna deteriorate considerably at great angles of inclination irrespective of the particular feeder system. These variations of the radiation pattern could be reduced by making the scanning sector of 120 = 35 + 85° symmetrical with respect to the in-phase case. This would require in practice an initial inclination of the antenna of 25° south which, at an overall length of 1 km, means that the north end of the antenna should be raised by 421 m. But even then the radiation pattern parameters at maximum angles of inclination would be those as given in Table 1 for θ_C = 60°. This proposition seems unrealizable, so that one must accept these variations in the radiation pattern and take them into account in actual operation.

When selecting the spacing between the individual elements of an equidistant scanned array, the principal requirement is the absence of diffraction peaks in the radiation pattern. As was shown in [3], the appearance of diffraction peaks is closely connected with the beam inclination angle θ_C and with the spacing between elements. Because of some design considerations, the spacing between the elements was chosen by us as being equal to d = 1.5 m, corresponding to d = $(\lambda_m/2)$ [λ_m = 3m]. It is evident that diffraction peaks may be expected on waves $\lambda \leq \lambda_m$ at $\theta_C \geq \theta_{C,max}$. The values of $\theta_{C,max}$ have been calculated, and the results are shown in Fig. 1. As follows from these results, in the short-wave portion of the frequency range (100 to

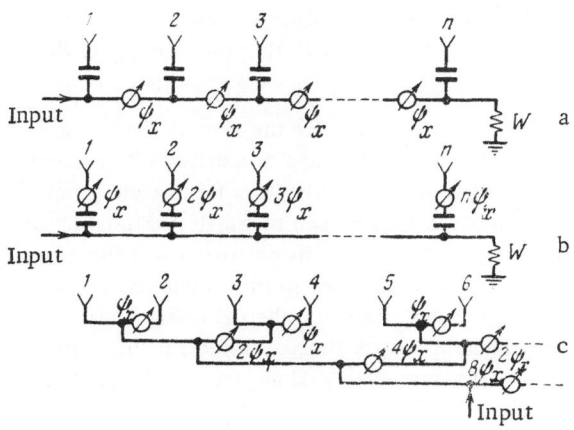

Fig. 2. Schematic connections of phase shifting elements.

120 Mc) it is impossible to incline the beam by more than $\theta_{C,max}$ without the appearance of a diffraction peak; this fact somewhat impairs the usefulness of the radio telescope at these frequencies.

Shunted wide-range dipoles placed at a distance of 90 cm from the counter-reflector are used as radiating elements of the North—South line exciter. The construction of these dipoles is the same as used in the East—West line (see [2], in which the principal characteristics of these dipoles are given).

In order to keep the direction of the principal maximum constant over the entire operating range (this is one of the most important demands that the exciter must satisfy), the phase shifter employed must provide in the elements phase lags which are proportional to the spatial delay $\psi = kd \sin \theta_C$ for the selected angle of inclination in the operating frequency range. A variable-length coaxial line with a characteristic impedance $W = 75 \, \Omega$ serves as such a phase shifter. The phase lag of this line is $\varphi = kl$. Provided that $\psi = \varphi$, i.e., $kl = kd \sin \theta_C$, we get $l = d \sin \theta_C$, i.e., the spatial delay is compensated by an appropriate length of the phase shifter irrespective of frequency.

Selecting the Electrical-Scanning Circuit

Introduction of phase lag into radiating elements of electrically scanned antennas is accomplished by means of phase shifting elements inserted into the feeder system. Three principal circuits of introducing phase lag may be cited (Fig. 2).

In the weakly coupled series circuit (Fig. 2a) [4], the phase lags of the radiating elements are identical and equal to ψ_x, and the necessary phase distribution in the radiating elements results from the fact that the radiated wave passes through all phase shifters in series.

In the weakly coupled parallel circuit (Fig. 2b), the phase lag is introduced in each element by a separate phase shifter whose phase lag increases in proportion to the radiator number. In the strongly coupled parallel circuit (Fig. 2c) [6], the total energy is applied to the radiating elements, the phase lag being provided by phase shifters introduced at the appropriate arms of the branching points.

Let us consider the principal features of all these circuits. The circuit in Fig. 2a has wide-range properties with respect to directionality, i.e., the angle of inclination remains constant within a range of frequencies. The total number of phase shifting elements required in this circuit is minimum as compared with the circuits in Figs. 2b and 2c. Systematic errors in amplitude—phase distribution due to mismatch between the radiating elements and the feeders supplying the energy are absent because of the equal electrical lengths from each dipole to the element coupling it to the main feeder line. The traveling wave mode in the main feeder determines the performance of the circuit and may be realized when the coupling between the main feeder and radiating elements is weak. This requires the dissipation of a large fraction of energy in the terminating load. Formation of a standing wave in the main feeder leads to increasing side lobes, because of the systematic errors in amplitude—phase distribution along the array. As has been shown in [7], the circuit of Fig. 2a requires the most accurate adjustment of phase shifters, since the random errors tend to accumulate when the necessary lag of each phase shifter is adjusted independently.

The random errors of inaccurate adjustment of phase shifters in the circuit of Fig. 2b do not accumulate and thus a lesser accuracy of adjustment may be tolerated. The circuit has wide-band properties with respect to the beam directionality if the phases are multiples of 2π, which requires large geometric dimensions of phase shifters if the latter are composed of variable-length lines. The performance of the circuit shown in Fig. 2b is determined mainly, as that in the case of circuit a, by the mode of the traveling wave in the main feeder.

If the radiators are mismatched, the systematic errors due to reflection result in increased side lobes [5]. These errors may be reduced by using matched directional couplers for connecting the radiating elements with the main feeder.

The adjusting accuracy and the number of phase shifting elements required for the circuit in Fig. 2c is half-way between those of the circuits of Figs. 2a and 2b. The circuit has wide-band properties with respect to directionality if no zoning occurs in the phase distribution (phases which are multiples of 2π at the given wavelength are not excluded). A cophase version of the circuit shown in Fig. 2c can be easily realized and, furthermore, it is quite simple to achieve inclination of the beam on both sides of its position when the array is excited in phase, which in the case of the circuit of Fig. 2a is quite complicated from a technical point of view. If the radiating elements of the array are well matched to the supply feeders, the necessary phase distribution in the radiators is automatically provided by inserting variable-length lines adjusted for the correct phase lag. Secondary reflection from the summing points may be eliminated by making the characteristic impedance of the summing feeder one half that of the feeder branches.

If the dipoles are not exactly matched to the feeder, multiple reflection from the dipoles and branching points will result in systematic errors in amplitude–phase distribution in the antenna. Let us calculate the ratio of voltages across unmatched loads connected in parallel through a matched T-junction. It may be easily shown that the voltage across the first load is

$$\dot{U}_1 = \dot{U}_0 \, \frac{1 + \dot{\rho}_{L_1}}{1 + \dot{\rho}_{L_1} e^{-i2kl_1}} \, e^{-ikl_1},$$ (3)

and that across the second load is, correspondingly,

$$\dot{U}_2 = \dot{U}_0 \, \frac{1 + \dot{\rho}_{L_2}}{1 + \dot{\rho}_{L_2} e^{-i2kl_2}} \, e^{-ikl_2},$$ (3')

where $\dot{U}_0$ is the voltage at the summing point, $\dot{\rho}_{L_1}$ and $\dot{\rho}_{L_2}$ are the reflection coefficients from the loads Z_1 and Z_2, respectively, and l_1 and l_2 are the corresponding lengths of feeders from the loads to the summing point.

If $\dot{\rho}_{L_1} = \dot{\rho}_{L_2} = \dot{\rho}_L$ ($Z_1 = Z_2$), and if $l_1 = l_2 + \Delta l = l + \Delta l$, the voltage ratio will be

$$\frac{\dot{U}_1}{\dot{U}_2} = \frac{1 + \dot{\rho}_L e^{-i2kl}}{1 + \dot{\rho}_L e^{-i2k(l+\Delta l)}} \, e^{-ik\Delta l}.$$ (4)

However, $\dot{\rho}_L e^{-i2kl} = \dot{\rho}$ is the load reflection coefficient referred to the summing point at $\Delta l = 0$ (in-phase case). Expression (4) may finally be written as

$$\frac{\dot{U}_1}{\dot{U}_2} = \frac{1 + \dot{\rho}}{e^{ik\Delta l} + \dot{\rho} e^{-ik\Delta l}}.$$ (5)

In case of matched loads, $\dot{\rho} = 0$, $\dot{U}_1/\dot{U}_2 = e^{-ik\Delta l}$, i.e., the phase difference between the voltages across Z_1 and Z_2 is equal to the difference between the electrical lengths of feeders 1 and 2. The analysis of expression (5) may be most conveniently made on the complex plane of the reflection coefficient $\dot{\rho}$. Using the method of conformal mapping, one can obtain the radius and the coordinates of the center of a circle on the complex plane $\dot{k} = \dot{U}_1/\dot{U}_2$, which are connected through expression (5) with the circle of radius $|\dot{\rho}|$ on the plane of the reflection coefficient.

The abscissa of the mapped circle center is

$$x_0 = \cos k\Delta l.$$ (6)

Its ordinate is

$$y_0 = -\frac{1 + |\dot{\rho}|^2}{1 - |\dot{\rho}|^2} \sin k\Delta l.$$ (7)

Table 2

$|\dot{\rho}|$	R_{max}	y_0	$\left|\dfrac{\dot{U_1}}{\dot{U_2}}\right|_{min}$	α_{max}
0.2	0.417	—1.08	0.665	±23°
0.4	0.950	—1.35	0.430	±44°
0.6	1.870	—2.12	0.250	±62°

The radius of the circle is

$$R = \frac{2\,|\dot{\rho}\,|}{1 - |\dot{\rho}\,|^2}\,\sin k\Delta l. \tag{8}$$

From expressions (6), (7), and (8) it follows that in the case $k\Delta l = \pi/2$ ($\Delta l = -\lambda/4$)

$$R = R_{max} = \frac{2\,|\dot{\rho}\,|}{1 - |\dot{\rho}\,|^2};\quad x_0 = 0;\quad y_0 = -\frac{1 + |\dot{\rho}\,|^2}{1 - |\dot{\rho}\,|^2}.$$

Let us estimate the maximum deviation of $|\dot{k}|$ from unity and of the voltage phase difference $\alpha = \varphi_1 - \varphi_2$ from $k\Delta l = 90°$ for various $|\dot{\rho}|$.

Circles representing the complex voltage ratio $\dot{k} = \dot{U_1}/\dot{U_2}$ are plotted in Fig. 3 according to the data of Table 2. It may be easily noted that

$$\left|\frac{\dot{U_1}}{\dot{U_2}}\right|_{min} = |y_0| - R = \frac{(1 - |\dot{\rho}\,|)^2}{1 - |\dot{\rho}\,|^2}$$

and

$$\alpha_{max} = \arcsin\frac{R}{|y_0|} = \arcsin\frac{2\,|\dot{\rho}\,|}{1 + |\dot{\rho}\,|^2}.$$

The calculated values of $|\dot{U_1}/\dot{U_2}|_{min}$ and α_{max} are also given in Table 2. As may be seen from the calculated results, unmatched loads lead to systematic errors in amplitude—phase distribution. To reduce these errors, a multiterminal network must be connected at the branching points, such that it decouples the dipoles one from the other and at the same time is matched to the dipole feeder.

A wide-band hybrid ring whose balanced version (two-wire version) was described in [8] may serve as such a multiterminal network. Such a ring has a traveling wave ratio (TWR) better than 0.8 with a bandwidth of 40 Mc as seen from arms 1 and 2 when matched loads are connected to arms 4 and 3, and provides a decoupling of about 30 dB between arms 1 and 2. We have constructed a balanced ring for a design frequency of 80 Mc, and plotted the amplitude and phase characteristics of arms 1 and 2 as a function of the arm length l_{z_1}. Arms 1 and 2 were loaded by impedances z_1 and z_2. The mismatch between the loads and feeder produced a TWR of 0.3, which corresponds to the lower limit of matching between the dipoles and feeder. Experiments proved that within the frequency band from 60 to 100 Mc the maximum systematic error did not exceed 10° in phase and 10% in amplitude. Such a result may be assumed quite satisfactory if it is taken into account that the dipoles have a TWR ≥ 0.3 within the indicated frequency band. However, the application of a two-wire version of hybrid rings presents difficulties associated with their construction as well as with their incorporation into the feeder system. We have thus developed a coaxial version of a wide-band hybrid ring with a π-bend. A detailed description of this ring and the experimental results obtained with it will be given below.

Estimation of the Effect of Errors on the Radiation Pattern of the Exciter

In the realization of the required amplitude—phase distribution in the array, random errors due to inaccurate adjustment of the phase shifting elements and to the spread of the feeder elements parameters tend to accumulate. It is well known that an increase of random errors in amplitude—phase distribution results in increased side lobes, reduction of the antenna gain, and in a deviation of the radiation pattern from the desired direction.

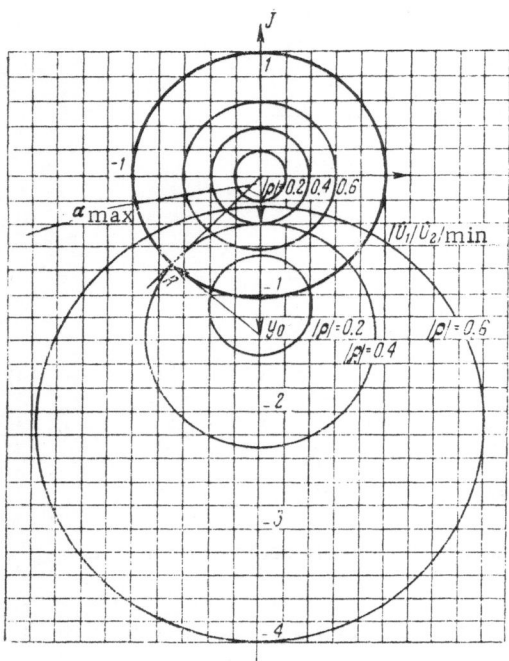

Fig. 3. Circular plots of the function $\dot{U}_1 / \dot{U}_2 = f(\dot{\rho}_1 k \Delta l)$.

The influence of amplitude and phase errors on the level of side lobes in a cross-type telescope has been considered in [4]. Assuming normal error distribution, the following expression has been derived for the rms error of the side-lobe level:

$$\{\overline{[E\,(\theta,\varphi)]^2}\}^{1/2} = \frac{[\sigma_{I_n}^2 + \sigma_{\Delta\psi_n}^2]^{1/2}}{\sqrt{N}}, \tag{9}$$

where N is the number of radiating elements, σ_{I_n} is the rms amplitude error, and $\sigma_{\Delta\psi_n}$ is the rms phase error. Thus, if the rms error in amplitude distribution is 20% and the rms error of phase distribution is 10°, the increase of side-lobe power level in an array consisting of 600 dipoles will amount to 1.3%. A detailed analysis of random errors in linear arrays has been made in [9]. An expression has been derived which relates the actual radiation pattern with an ideal one:

$$\overline{P\,(U)} = e^{-\sigma_\delta^2} P_d\,(U), \tag{10}$$

where $\overline{P(U)}$ is the mathematic expectation of the pattern, σ_δ^2 is the phase dispersion, and $P_d(U)$ is the ideal pattern. Thus, with an rms phase error of 10°, the reduction of power gain will be 3%. The discrepancy between the increase of side lobes and reduction of gain may be explained by broadening of the principal lobe of the pattern.

The same work also cites an expression which connects the rms error in phase distribution with the desired degree of accuracy of directivity of the radiation pattern:

$$\sigma_\delta\,(P,\gamma) = \frac{\pi}{a_p}\,\frac{\gamma}{\gamma_0}\sqrt{\frac{N}{\sigma}}, \tag{11}$$

where N is the number of elements in the array, a_p is a parameter which takes into account the probability of the pattern being within the allowed limits of deviation (if P = 0.99, a_p = 2.58), and γ / γ_0 is the permissible deviation from the desired direction which is fulfilled with a probability P. Thus, if the deviation of a 10 ft-wide radiation pattern is not to exceed $\frac{1}{10}$ of its width with a probability of P = 0.99, we have

$$\sigma_\delta\,(0.99;\,1') = 1.21\,\text{rad}\quad (\sigma_\delta = 70°)\quad \text{for}\quad N = 600.$$

If the scanned pattern belongs to a fixed line with cophased dipoles, we have

$$\sigma_\delta\,(0.99;\,1') = 0.87\,\text{rad}\quad (\sigma_\delta = 50°)\quad \text{for}\quad N = 300.$$

If the angle of scan is increased, the tolerable deviation from the given phase distribution should be reduced by $\cos\theta_C$, where θ_C is the angle of beam inclination,

$$\sigma_{\delta C} = \cos\theta_C \cdot \sigma_\delta. \tag{12}$$

Thus, for an angle of inclination θ_C = 60°, the permissible phase deviation reduces to 25°. Amplitude errors produce effects of second order of smallness and were neglected.

The above evaluation of the effects of amplitude—phase distribution errors on the radiation pattern indicates that rms phase errors of ±10° and rms amplitude errors of ±20% do not affect the radiation pattern significantly. The tolerances indicated above have been accepted as the tuning accuracy of the antenna and feeder system.

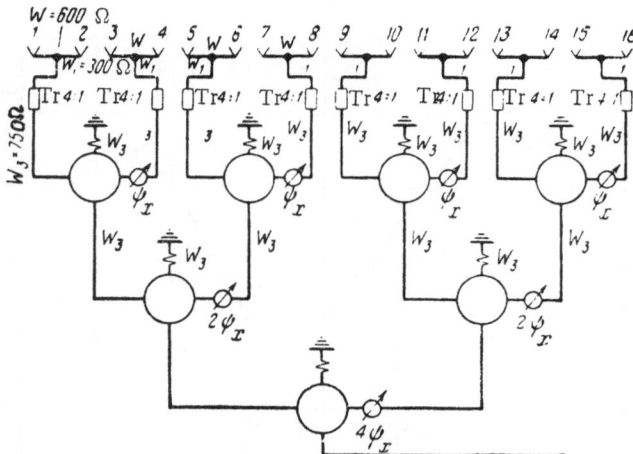

Fig. 4. Circuit diagram of one 16-dipole section of the exciter.

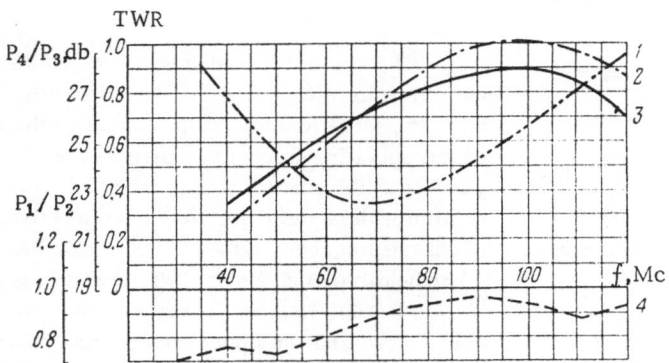

Fig. 5. Characteristics of coaxial hybrid ring with π-bend ($f_P = 100$ Mc): P_4/P_3 curve (1), calculated TWR curve (2), experimental TWR curve (3), and P_1/P_2 curve (4).

Electrical Circuit of the North — South Exciter

The electrical circuit of the exciter of one 16-dipole span of the KR-1000 North—South line is shown in Fig. 4.

A pair of dipoles is connected together by a 600-Ω two-wire feeder which, in turn, is connected through a balanced flat 300-Ω cable to a wide-band 4:1 balancing impedance transformer. RK-3 cables lead down to the railway where all summing and phase shifting elements of all individual spans are located. The high-frequency energy from each antenna array is led by means of a 500-m long RKK-5/18 cable into the switching basement room, where further summing of all the North—South line energy takes place. Inclination of the antenna directivity pattern is realized by connecting variable-length lines into the appropriate branches of the feeder system.

Elements of the Exciter Feeder System

Efficient electrical scanning of the exciter within a wide range of frequencies requires that all elements of the feeder system have good wide-band properties. The principal elements of the feeder system are the hybrid rings (decoupling element) and the 4:1 balancing impedance transformers (300-75 Ω) which couple the balanced dipoles to the coaxial feeder system.

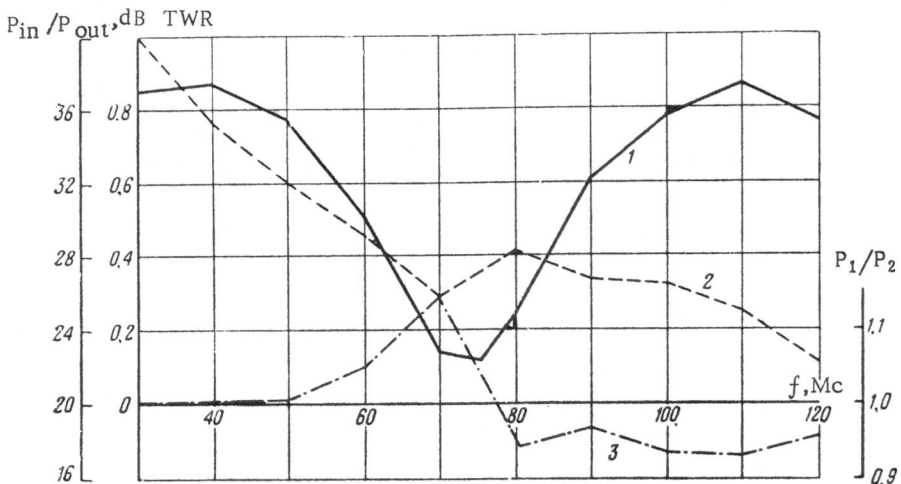

Fig. 6. Characteristics of a hybrid ring with a π-bend (f_P = 36 and 108 Mc):
TWR curve (1), P_{in}/P_{out} curve (2), and P_1/P_2 curve (3).

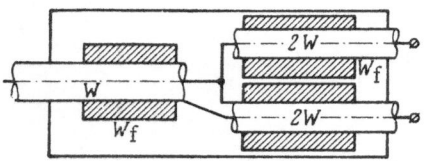

Fig. 7. Balancing 4:1 transformer.

The coaxial hybrid ring is a conventional bridge whose opposite arms are effectively decoupled (better than 22.5 dB). Depending upon the specific demands presented to the exciter at various regions of the operating frequency range, several versions of a hybrid ring are possible within the frequency range of 30 to 120 Mc. Thus, for example, the hybrid ring may have a resonant frequency at the center of the frequency range. Such a ring has been investigated. It has a bandwidth of ±25% at a TWR level ≥0.8, and ±50% at a TWR ≥ 0.5 (Fig. 5). However, the normal operation of the KR-1000 requires stressing of the lower region of frequencies (from 30 to 55 Mc), where the TWR of dipoles is low, and the upper region (from 90 to 120 Mc), where the signal intensity is low. For this reason, a hybrid ring with a resonant frequency f_{P_1} = 36 Mc ($\lambda_{P_1}/4$ = 2.08 m) is more suitable. Such a ring has a second resonance at $\frac{3}{4}\lambda_{P_2}$ = 2.08 m (f_{P_2} = 108 Mc). The TWR of a hybrid ring with f_P = 108 and 36 Mc is shown in Fig. 6 as a function of frequency.

The necessary decoupling of opposite arms is achieved by using a wide-band device which gives an additional phase shift of 180° in one of the bridge arms (π-bend).

The frequency dependence of decoupling provided by the rings is shown in Fig. 5,1 and in Fig. 6,2. Reduction of iterative attenuation at frequencies other than resonant may be explained by impaired matching of the ring at these frequencies. Balanced construction of the ring ensures independent division of power among arms 1 and 2 with an accuracy ±10%.

Various possible designs of π-bends and 4:1 transformers have been described in [10, 11]. We have constructed the transformer shown in Fig. 7. The balancing 4:1 impedance transformer is used for feeding the balanced dipoles of the exciter within the frequency band from 30 to 120 Mc. Balancing of the transformer output voltage is characterized by phase unbalance $\psi = 2\pi d/\lambda$ (wave sloping), where d is the distance between the voltage nodes on the two wires of the balance feeder, and also by amplitude unbalance

$$\delta\% = \frac{|\dot{U}_1| - |\dot{U}_2|}{|\dot{U}_1| + |\dot{U}_2|},$$

where $|\dot{U}_1|$ and $|\dot{U}_2|$ are the antinode voltages on the corresponding feeder wires. It may be seen in Fig. 8 that the constructed transformers have an amplitude unbalance $\delta \leq 4\%$ and a phase unbalance $\psi \leq 4°$. The indicated maximum unbalances are determined principally by the accuracy of transformer construction. The TWR of the transformers over the entire frequency range is between 0.8 and 0.85. The insertion loss is about 0.85 dB over the entire range.

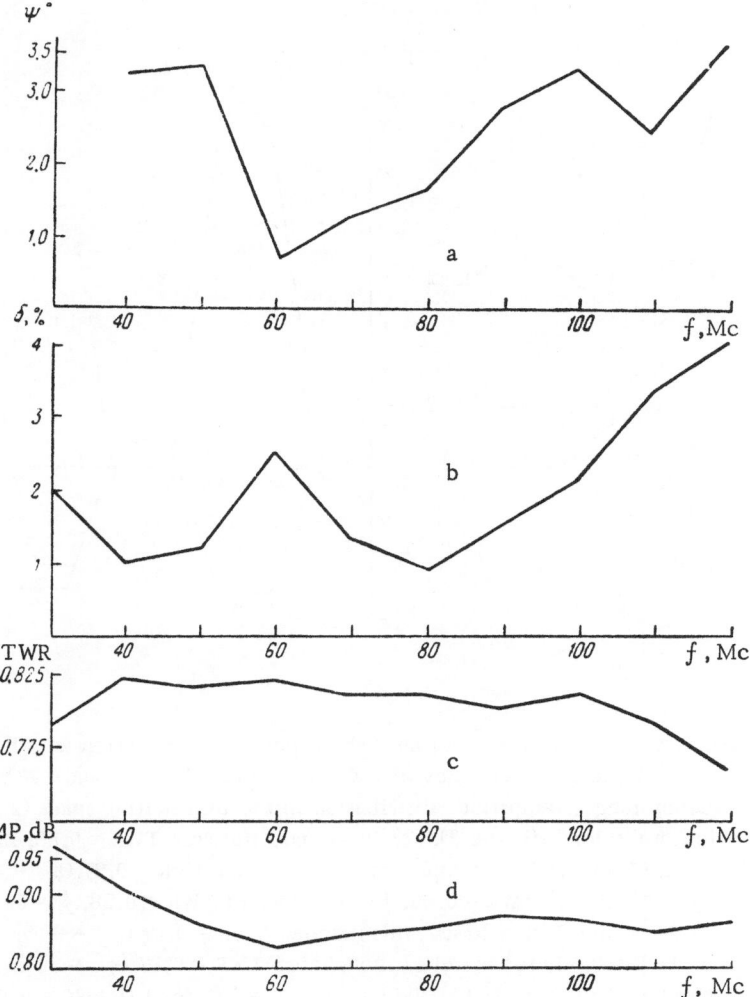

Fig. 8. Experimental characteristics of the balancing transformer within the frequency range: phase unbalance (1), amplitude unbalance (2), TWR (3), and insertion loss (4).

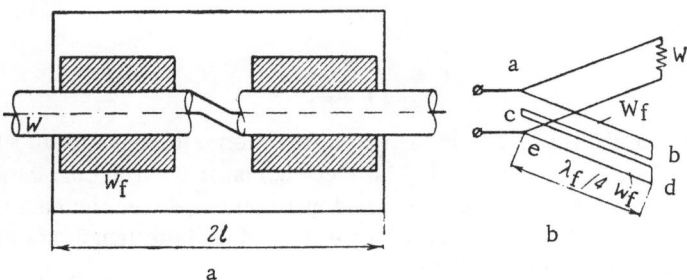

Fig. 9. π-Bend (a) and its equivalent circuit (b).

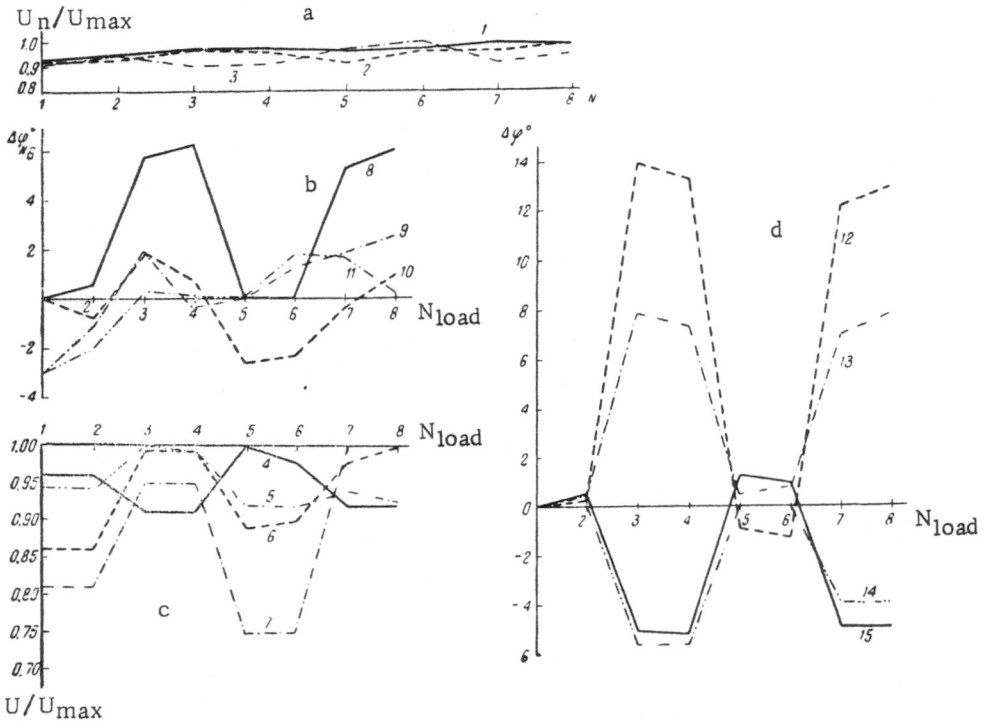

Fig. 10. Experimental characteristics of an eight-dipole exciter section with hybrid rings. Left: $f_p = 100$ Mc at a frequency of 110 Mc. Right: $f_p = 36$ and 108 Mc within the frequency range: amplitude distribution in the terminating loads (a and c), phase distribution in the loads (b and d). 1) Cophased, matched TWR = 0.86; 2) cophased, mismatched TWR = 0.27; 3) nonphased, mismatched TWR = 0.27 ($\psi_x = 90°$); 4) TWR = 0.45, $f = 45$ Mc; 5) TWR = 0.60, $f = 110$ Mc; 6) TWR = 0.28, $f = 55$ Mc; 7) TWR = 0.65, $f = 90$ Mc; 8) nonphased, mismatched TWR = 0.27 ($\psi_x = 90°$); 9) cophased, matched TWR = 0.86; 10) matched, nonphased TWR = 0.87 ($\psi_x = 90°$); 11) cophased, mismatched TWR = 0.27; 12) TWR = 0.28; $f = 55$ Mc; 13) TWR = 0.65, $f = 90$ Mc; 14) TWR = 0.60, $f = 110$ Mc; 15) TWR = 0.45, $f = 45$ Mc.

An important component of the wide-band elements of the feeder system is the π-bend. Theoretical principles of operation of a π-bend with ferrites are given in [10]. To ensure wide-band properties of the π-bend, the chosen type of ferrite should provide a proportional variation of parameters μ_r and ε_r with frequency, so that the relationship

$$l = \frac{\lambda}{4\sqrt{\mu_r \varepsilon_r}} , \qquad (13)$$

holds (within the frequency range). Here, l is the length of the short-circuited sleeve filled with ferrite, μ_r is the magnetic permeability of the ferrite, and ε_r is its dielectric constant. The coaxial π-bend (Fig. 9) has been made of RK-3 coaxial cable connected core-to-braid and braid-to-core, and enclosed in a solid shield. To prevent currents from flowing in the cable braid, ferrite rings of definite length are put on the braid and form a short-circuited sleeve with an input impedance

$$Z_{in} = W_f \tanh \gamma l, \qquad (14)$$

where W_f is the characteristic impedance of the coaxial line partly or fully filled with ferrite.

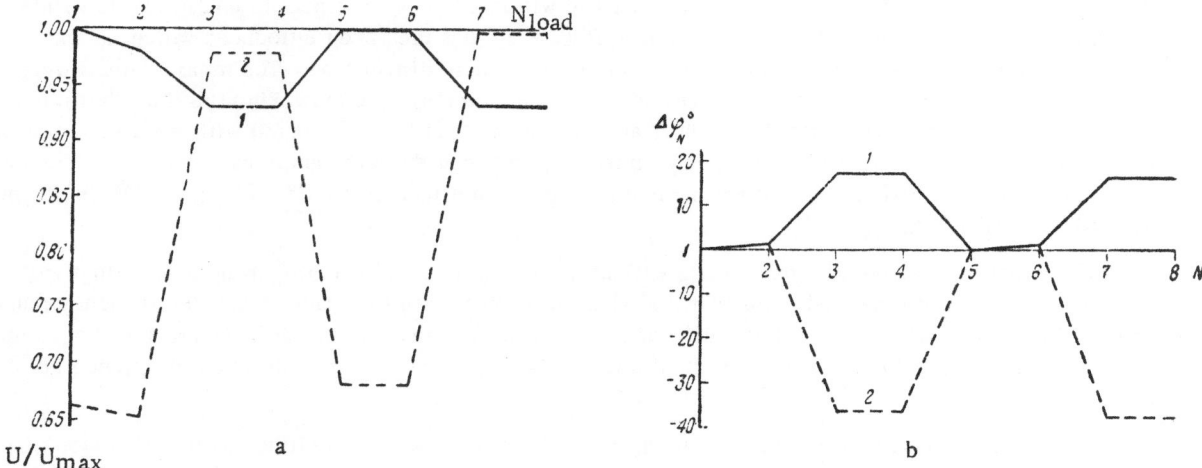

Fig. 11. Amplitude—phase distribution in loads with TWR = 0.27 for an exciter section at f = 50 Mc. Hybrid rings with f_P = 100 Mc. Amplitude distribution (a), phase distribution (b) (deviation from the linear relationship with ψ_X = 90°). 1) With hybrid rings; 2) without hybrid rings.

The basic calculations and analyses were given in [10]. Type F-600 ferrite has been used in the π-bend constructed by us. The properties and characteristics of this ferrite have been estimated according to [12]. The chosen ferrite gives a quite good constancy of the relationship (13) in the frequency range from 30 to 120 Mc, providing a voltage phase shift of 180° ± 2°. It has been found experimentally that the deviations of phase from 180° are within the measuring error. The constructed π-bend provides close matching (TWR = 0.86 to 0.92) over the entire frequency range, its insertion loss being P_{out}/P_{in} ≤ (0.7-0.8) dB. An analysis of the equivalent circuit (Fig. 9b) shows that the increase of input impedance of the ferrite-filled line reduces the shunting effect of the short-circuited sleeve.

Thus, elements of the exciter feeder system have been developed which are efficient enough for incorporation into the electrical scanning circuit of the North—South line.

Results of Experimental Investigation of an Exciter Section

Experimental results were obtained for an exciter section consisting of eight dipoles. Eight identical non-radiating loads were connected to the feeder in place of the dipoles for measuring the amplitude—phase distribution for various introduced phase lags. The measurements were made with dummy dipoles, since with real dipoles it is difficult to isolate the systematic errors because of edge effects which result in the dipoles having different input impedances depending on their location in the exciter. The latter results in variations of the amplitude—phase distribution which are difficult to account for. The terminating loads had identical parameters. All necessary experiments were conducted with two types of hybrid rings: rings with f_P = 100 Mc, and rings with f_P = 36 and 108 Mc.

The amplitude and phase characteristics of the loads were plotted at 110 Mc using hybrid rings with f_P = 100 Mc in four different cases:

1. Matched loads, cophased version ψ_X = 0°.

2. Mismatched loads, cophased version ψ_X = 0°.

3. Matched loads, nonphased version ψ_X = 90°.

4. Mismatched loads, nonphased version ψ_X = 90°.

The cophased version has been investigated in order to estimate the amplitude and phase errors which occur in the exciter because of the spread of parameters of the individual feeder elements. It was found by measurement that a cophased exciter section has been tuned to within $\pm 3.0°$ with respect to phase, and to within $\pm 4\%$ with respect to amplitude (Fig. 10). A phase shift $\psi_X = 90°$ has then been introduced which, as was shown above, causes maximum systematic errors in amplitude−phase distribution. The measurements were conducted with loads having a TWR of 0.86 and 0.27. The characteristics plotted in Fig. 10b show that the phase distribution errors for loads with TWR = 0.27 are not greater than 6° as compared with the distribution obtained with loads having a TWR = 0.86, i.e., large mismatch between the loads and feeder does not affect significantly the phase distribution in circuits using hybrid rings. Amplitude errors did not exceed 10% in all the investigated cases (Fig. 10a).

Similar experiments have also been made with an eight-dipole exciter section using hybrid rings with $f_P = 36$ and 108 Mc. The measured amplitude and phase distributions in the loads within the frequency range are shown in Figs. 10c and 10d. The phase shift introduced at each frequency was 90°. Figure 10d shows the phase deviation of currents in the loads from the desired linear relationship with a 90° phase difference between loads.

An analysis of the results shows that maximum deviation from unity in amplitude distribution took place at 90 Mc, and that in phase distribution at 55 Mc. At these two cutoff frequencies of the hybrid ring, the latter's TWR is equal to 0.5.

The results obtained from measurements at 50 Mc are shown in Fig. 11. After the hybrid rings have been inserted into the branching points, the maximum phase deviations did not exceed 18° and the amplitude deviations 10%. It may be indicated, for comparison, that measurements made with an exciter section without hybrid rings but with the necessary matching conditions at branching points preserved gave, for the same phase lag, systematic phase and amplitude deviations of 36° and 35%, respectively. This once more proves the effectiveness of hybrid rings in an exciter whose radiation pattern is scanned within a range of frequencies, the matching between dipoles and feeder being imperfect.

The cited calculations and the tenative experimental results prove the efficacy of using hybrid rings as summing elements in the strongly coupled parallel circuit. It should be noted that in the discussed circuit a fraction of energy of the order $|\dot{\rho}|^2$ ($|\dot{\rho}|$ being the modulus of the dipole reflection coefficient) is dissipated in the rings, whereas in weakly coupled circuits not less than 50% of the energy is always absorbed in the terminating load. Furthermore, in weakly coupled circuits the noise of the terminating load enters the receiver input, impairing its signal-to-noise ratio. In the proposed circuit with hybrid rings the absorbing loads are efficiently decoupled from the receiver input and do not affect its signal-to-noise ratio.

The authors wish to express their gratitude to Doctor of Physicomathematical Sciences V. V. Vitkevich, who directed the scientific work for the construction of the KR-1000 radio telescope, and to Junior Scientific Associate Yu. I. Alekseev for his participation in the discussion of results.

Literature Cited

1. V. V. Vitkevich, Vestn. Akad. Nauk SSSR, No. 5 (1961).
2. Yu. P. Ilyasov and A. D. Kuz'min, this volume, p. 7.
3. W. H. Von Aulock, Proc. IRE, 48:1715-1727 (Oct., 1960).
4. B. Y. Mills, A. G. Little, K. V. Sheridan, and O. B. Slee, Proc. IRE, No. 1:67-84 (1958).
5. Z. A. Kurtz and R. S. Elliott, IRE Trans., Antennas and Propagation, AP-4, No. 4 (1956).
6. W. N. Christiansen, N. R. Labrum, U. R. McAlister, and D. S. Mathewson, Proc. IRE, B-108, No. 37:48-58 (1961).
7. O. G. Vendik, Izv. Vuzov. Radiotekhnika, No. 2:179-189 (1962).
8. W. Tyminsky and A. Hylas, Proc. IRE, No. 1: 31 (1953).
9. M. Leichter, IRE Trans., Antennas and Propagation, AP-8, No. 6:268-275 (1960).
10. T. R. O'Meara and R. L. Sydnor, Proc. IRE, 46(11):1848-1860 (1958).
11. Osima, Proc. IRE, 47 (5):998-999 (1959).
12. L. I. Rabkin, S. A. Soskin, and B. Sh. Epshtein, Ferrite Engineering, Gosénergoizdat (1962).

POSSIBLE METHODS OF BUILDING
LARGE RADIO TELESCOPES

V. V. Vitkevich and P. D. Kalachev

The design and fabrication of highly directional antennas for radio telescopes is one of the more complex problems associated with the development of radio astronomy.

The effort to find the simplest and most economical solutions in building large antennas has resulted in the appearance of a great number of different structural arrangements.

1. Fixed horizontal antennas with electrical beam steering.
2. Depressions in the ground, lined with plates or mesh and having a fixed beam.
3. Depressions in the ground with a movable secondary reflector.
4. Cruciform antennas of the "Mills Cross" type having limited movement.
5. Interferometers having a fixed and a variable base.
6. Fan antenna or an antenna of variable shape.
7. Classical, completely rotatable parabolic reflector.

All these antennas (items 1-6) have well-known technical and economic advantages and inherent specific disadvantages: small scanning angles, complicated phasing, complicated illumination, imperfect directional patterns, etc.

The classical antenna in the form of a completely rotatable parabolic reflector with its unlimited scanning angles, comparatively simple exciter, broad range, the possibility of operating simultaneously at several frequencies, etc., is undoubtedly the most acceptable to radio astronomers.

Considering the fact that questions of both design and technical fabrication of parablic reflectors are far from being settled, we will consider in our article only completely rotatable parabolic reflectors.

In the radio astronomy laboratory of the P. N. Lebedev Physics Institute, two possible methods for improving the construction of parabolic antennas have been considered and are being investigated at the present time, one for short waves and one new parabolic reflector scheme for long waves.

The greatest problem in designing a parabolic reflector is to provide maximum rigidity with minimum weight. In order to increase the rigidity of a reflector in all positions, and especially in the vertical, a new structural scheme is proposed for a parabolic reflector having strong bracings in the plane of the reflector aperture (inside the equipment) in the form of a cruciform truss * (Fig. 1).

A cruciform truss, which is of rigid welded construction, is first used as a production and checking template. For this purpose, the blade of the template, which is a sword-shaped band of steel checked on a counter-template, is made fast to the lower region of the truss, which adjoins the front of the reflector.

---

*In 1961, an Author's Certificate was issued to the authors of this article for an invention relating to the use of various design possibilities to increase the rigidities of parabolic reflectors, No. 134723, entitled "Parabolic Reflector."

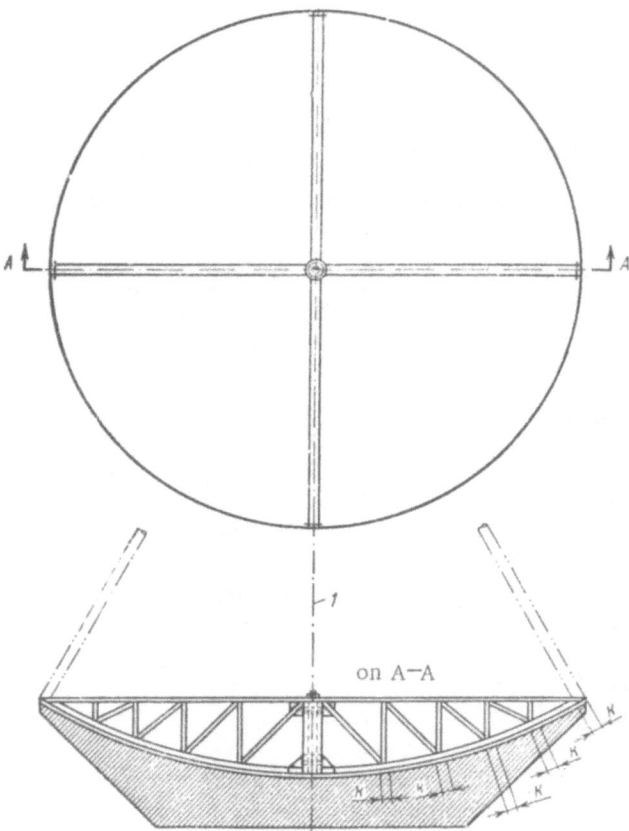

Fig. 1. Structural arrangement of a reflector having a strong
cruciform truss. 1) Geometric axis of the reflector and the
axis of rotation for the cruciform truss-template. K) Places
where the cruciform truss is joined to the strong reflector
frame.

In the center of the cruciform truss (at the crosspiece) there is a pedestal which supports radial and
thrust bearings by means of which the truss is mounted on a vertical column and can be rotated on it. The
vertical column is rigidly attached to the strong reflector frame.

In order to increase the surface of operations, four arms of cruciform shape can be employed simultaneously; the uniformity of the template blades and their installation on the truss is assured by the countertemplate.
After fabricating and checking the reflector, the template blades are removed and the lower joints of the strong
cruciform truss are fastened to the joints of the load-carrying reflector frame.

With sufficiently rigid fastenings between the joints of the cruciform truss and the strong reflector frame,
the rigidity of the latter is increased substantially.

In the presence of the strong cruciform truss reflector, shadowing is not only not increased, but is even reduced. The fact is that the four-rod design which carries the primary element cannot be attached to the reflector at its edges because of the hazard of large deformations in the reflector; as a result, the upper ends of
the rods will shade the primary element. This shadowing can be as much as 12% of the reflector area.

With the crosspiece, the ends of the rods that carry the primary element can be placed at the very edges
of the reflector on the ends of the cruciform truss without risk of getting appreciable reflector deformations.

The reflector shading by the cruciform truss itself is not over 3%. As for interference between the strong
cruciform truss and the reflector due to the reflected waves, this effect, in the opinion of the authors, intro-

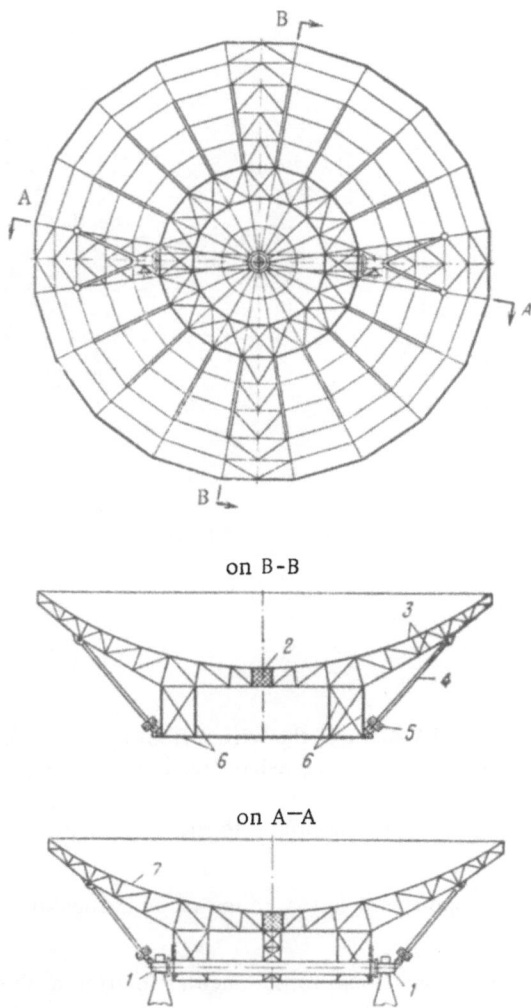

on B-B

on A-A

Fig. 2. Structural arrangement of a reflector having controlled supporting braces. 1) Supporting journals of the reflector; 2) central bushing; 3) radial truss; 4) support brace; 5) unit for controlling the length of the brace; 6) three-dimensional ring truss; 7) parabola.

duces no appreciable distortion of the electric characteristics. Absorptive coatings must play an essential role in this problem.

The independent use of a crosspiece is limited by reflector dimensions (to a diameter in the order of 40 m). In order to build a larger sized parabolic reflector (more than 40 m), more radical design methods are required.

In the radio astronomy laboratory of the P. N. Lebedev Physics Institute, a new structural arrangement for a parabolic reflector is being studied in which maintenance of the correct parabolic shape during its rotation throughout the angle of elevation is ensured by automatically controlling the length of the main support braces.

Inasmuch as the primary and constantly acting load on a reflector is its own weight, first consideration was given to a structural arrangement which fulfills the requirement of maintaining a constant shape under the influence of the weight loading. A brief description of the proposed design is presented below.

The strong frame of the reflector consists of: 1) a three-dimensional ring truss in the form of an icosagon; 2) a central bushing; 3) radial parabolic trusses, rigidly joined to the central bushing; 4) chord trusses; 5) diagonal braces located in the lower panels of the reflector frame; 6) strong supports, which are adjustable in length; 7) a pin hoop; 8) a vertical tetrahedron truss joining the central bushing with the pin hoop; 9) two bracing trusses joining the lower point of the tetrahedron truss and the pin hoop with the two reflector supports.

Each radial parabolic truss can be treated like a truss that is lying on three supports (disregarding the ring point of the radial truss which is joined to the peripheral frame chord truss). The central bushing and the three-dimensional ring truss can be treated like fixed supports of the radial trusses, and the three support points are the strong supporting braces which are controlled in length; consequently, these supporting points are movable.

While set up and checked in the horizontal position, the reflector includes deformations due to its own structural weight which are in this case in a latent form. In the horizontal position of the reflector, the weight loadings act on it symmetrically. When the reflector is rotated to an angle of elevation, the weight loadings are changing from symmetrical to skew-symmetrical; in the vertical position they become skew-symmetrical. In this case, the strong radial trusses will have different deformations depending on the location of the radial truss. The lower trusses located directly on the vertical diametral cross section will have deformations equal to the differences between the deformations at the vertical and horizontal reflector positions. The upper trusses will have deformations equal to the sum of the corresponding deformations. The radial trusses located on the horizontal diametral cross section will have deformations equal to the differences of the deformations: from the symmetrical loadings (horizontal reflector position) and from the deformations at the ends of the

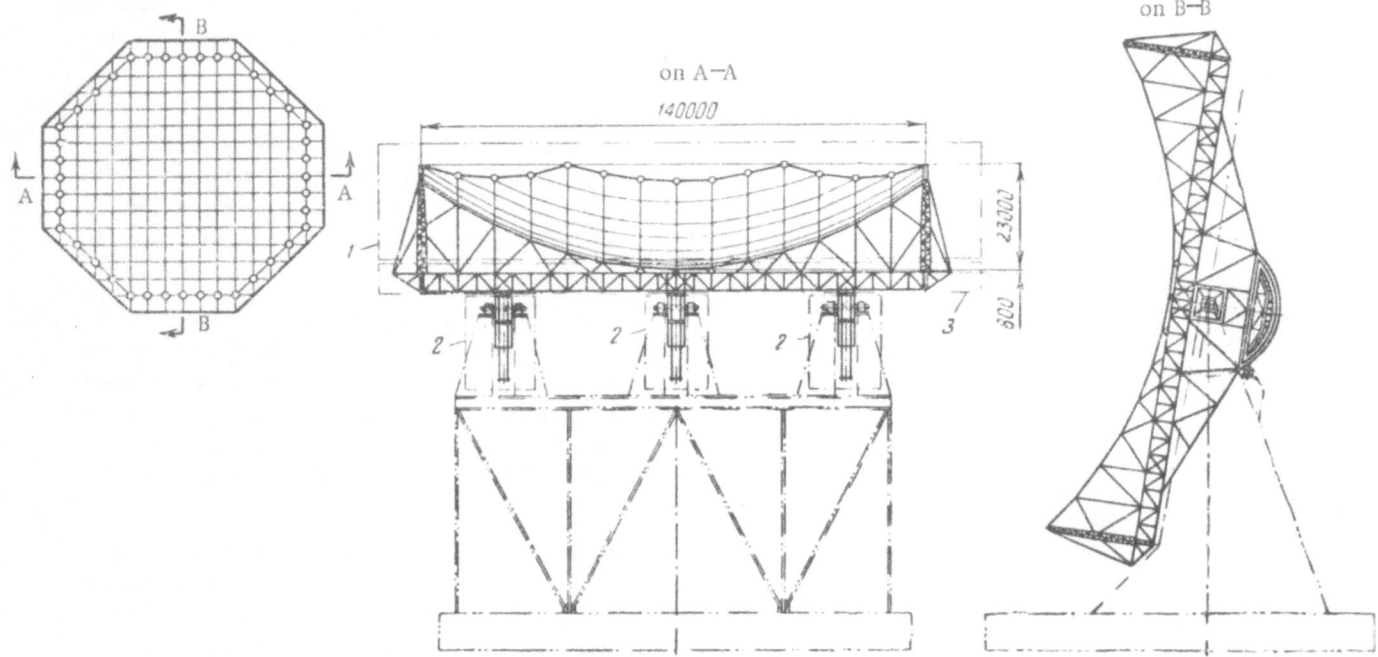

Fig. 3. Structural arrangement of a reflector having flexible belts. 1) Weight of the wire reflecting system 55 m; 2) weight of the three sectors with reinforcements, 195 m; 3) weight of the plane-parallel truss, 850 m.

journals supporting the reflector, which are imparted by means of the supporting braces to the arms of the corresponding radial trusses.

The deformations of all the intermediate radial trusses are proportional to the cosine of their angular inclination to the diametral cross section.

To a first approximation, all the deformations considered are proportional to the angular rotation of the reflector, reckoned from the zenith.

The lower ends of the strong supporting braces are fastened to the lower joints of the load-carrying, three-dimensional ring truss by means of a mechanical screw jack having an electrically controlled motor. In Fig. 2, the reflector is shown in a plan view, and two diametral sections through A—A and B—B are given. By virtue of the available calculated deformations, automatic control of the supporting brace lengths can be carried out in accordance with a specified program to compensate for these deformations. In this instance, of course, only the deformations due to the reflector's own structural weight are compensated.

Inasmuch as the complex deformation from wind loads requires much more complicated arrangements, it is advisable, for intermediate sizes of reflectors (up to 50 or 60 m), to restrict the compensation to deformations from the inherent weight. There will be some loss of useful operating time for the radio telescope when the wind velocity exceeds v = 9 m/sec. This loss of time is not over 12%.

The existence of temperature deformations is also going to result in some loss of radio-telescope operating time.

It is quite evident that in the future the problem of automatically controlling a reflector frame in order to maintain the correct parabolic shape will have to be solved compositely, taking into account wind loads and the effects of temperature.

At the present time, this question is being investigated in the radio astronomy laboratory of the P. N. Lebedev Physics Institute, and a more refined structural arrangement for suspending a reflector on a rotatable bearing mechanism is being developed.

It should be noted that the reflector suspension scheme, or the way in which it is fastened to the rotatable bearing mechanism, is the decisive condition for getting the minimum deformations.

In this regard, the 76-m "Jodrell Bank" reflector in England, which was made in the form of a three-support beam, cannot be considered successful.

*   *   *

For long-wave operation, starting at 1.5 m and longer, a new structural arrangement for a movable parabolic reflector having flexible belts (Fig. 3) has been developed in the radio astronomy laboratory of the P. N. Lebedev Physics Institute. In this arrangement, the load-carrying frame and the reflecting system are entirely separate. The load-carrying frame of the reflector is in the form of a three-dimensional plane-parallel truss, reinforced by braces which are joined to a truss having serrated (pin) sectors. The number of reflector supports that is structurally permissible is more than two, in our case being three. Correspondingly, this has three serrated sectors and three drive mechanisms, which are coupled mechanically or electrically to shafts.

The flat load-carrying truss in the plan view has the shape of a regular octagon with tensioning columns located around its outline.

Flexible wire belts which form a large grid pass out horizontally and vertically from the tops of each opposed pair of columns. The assembly of the reflector is carried out in the horizontal position of the flat load-carrying truss.

The flexible belt stretched between the tops of the corresponding columns assumes the natural shape of a catenary, which is close to a parabola. Steel wire braces are attached to points on a belt which are located at distances equal to the spacing between adjacent columns. By means of these wire braces, the flexible belt is pulled in toward the load-carrying truss in order to obtain the necessary shape of the flexible belt, which is simply obtained as a broken line. The closeness of the junction points is determined by the permissible deviation from the prescribed parabola. Flexible belts braced by this method maintain the shape thus obtained for any position of the reflector in space.

Thin wires, which constitute the fine reflecting network, are attached to the grid of flexible belts; the dimensions of the cells in the fine network are dictated by the received wavelength. For instance, with a reflector 140 × 140 m in size and an operating wavelength of $\lambda = 4$ m, the separation between the flexible belts (columns) is 10 m and a cell of the fine network is 250 × 250 mm.

The proposed design for a parabolic reflector combines within itself the simplicity of fabricating a plane-parallel, three-dimensional, load-carrying truss and the lightness of a wire reflecting system.

A PARABOLIC REFLECTOR WITH SCREENS

V. V. Vitkevich and P. D. Kalachev

Parabolic antennas are in current use in various branches of radio engineering, radiophysics, and radio astronomy. Their development has progressed toward the design of increasingly large structures capable of observing at wavelengths as short as possible.

In several countries there are radio telescopes 25 m in diameter [1], operating efficiently to wavelengths as short as 3 cm. At the Lebedev Physics Institute, a radio telescope 22 m in diameter has been developed [2] that can operate to 1-cm wavelength. In the United States, a radio telescope has recently been built that has a parabolic reflector 37 m in diameter and apparently is capable of operating to wavelength $\lambda = 1$ cm; and a parabolic reflector about 40 m in diameter has been designed with a minimum wavelength of about 3 cm.

The larger-diameter parabolic antennas perform two functions: because of their large area they ensure high reception sensitivity, and they furnish an instrument of high resolving power.

If we have an antenna system with radio receiving equipment connected to it, the basic sensitivity of the system will in the final analysis be governed by the noise that arrives at the input of the radio receiver. We shall denote the corresponding noise temperature at a given wavelength by T_n, so that the noise energy at the input will be $kT_n\Delta f$.

The total noise temperature T_n (and hence the noise energy) consists of antenna noise (including noise in the transmission channels) and receiver noise, or

$$T_n = T_{n,a} + T_{n,r}.$$

The value of $T_{n,r}$ depends on the type of receiving equipment, on the quality of its tuning, and on other parameters (the bandwidth, the gain, etc.). It is noteworthy that engineering developments have led to increasingly small values of $T_{n,r}$. At the present time, parametric and paramagnetic systems are being designed for centimeter and decimeter wavelengths with a noise temperature of the order of a few tens of degrees, and systems are in prospect with a noise of only a few degrees. In coming years we may anticipate wide application of equipment that can operate in the range from millimeter to centimeter and decimeter wavelengths with an internal noise temperature less than 10°K.

The quantity $T_{n,a}$ consists essentially of four components: the noise of galactic radio emission, $T_{n,G}$; atmospheric noise, $T_{n,A}$; the noise in the feeder systems, $T_{n,f}$; and noise emanating from the earth, $T_{n,E}$:

$$T_{n,a} = T_{n,G} + T_{n,A} + T_{n,f} + T_{n,E}$$

The noise from the galaxy (the general background of cosmic radio waves) depends on the region toward which the antenna is pointed and also on the wavelength. At meter wavelengths, $T_{n,G}$ is very large and can reach values of several hundred or several thousand degrees; but as the wavelength becomes shorter, $T_{n,G}$ declines sharply. At wavelengths from a few centimeters and up to 10 cm (we shall be interested primarily in this wavelength range), the value of $T_{n,G}$ over a large part of the celestial sphere is no more than 10-20°K.

The atmospheric temperature $T_{n,A}$ is small for meter, decimeter, and centimeter wavelengths; for attitudes of 5-10° and more it is equal to 3-5°K. At wavelengths of 1-2 cm and at small attitudes, this quantity may have a somewhat higher value.

The temperature $T_{n,f}$ depends on the loss in the lines transmitting energy from the focus to the receiver input. If the first stage of the high-frequency amplifier is placed near the focus, $T_{n,f}$ can normally be held to 3-5°K.

Finally, $T_{n,E}$ depends on the temperature of the ground at the given wavelength and on the amount of irradiation of the earth. At centimeter and decimeter wavelengths, $T_E \approx 200$°K. If Ω_E denotes the effective solid angle of exposure of the earth to the reflector feed, and F_E the corresponding relative value for the intensity of the portion of the feed pattern through which radiation from the earth is admitted, then

$$T_{n,E} \approx \frac{\Omega_E F_E T_E}{\Omega_0} = \alpha T_E,$$

where Ω_0 is the effective solid angle of exposure to the reflector feed. In practice, we may have $\alpha \approx \frac{1}{3}-\frac{1}{5}$, so that $T_{n,E} \approx 70-40$°K.

This value may be smaller by a factor of about 1.5-2 for short-focus reflectors, but it probably cannot be reduced much below 20°K.

With reference to the wavelength range from 1 to 10-20 cm, we may state the following points:

a) In the near future we can expect that in many cases the inequality $T_{n,r} < T_{n,a}$ will be satisfied; that is, antenna noise rather than noise in the receiving equipment will limit the sensitivity of the system.

b) Of all the components of antenna noise, the noise from the earth has the greatest value; in fact, $T_{n,E} > T_{n,G} + T_{n,A} + T_{n,f}$. The problem therefore arises of substantially reducing the noise from earth sources.

The suggestion we are about to offer is intended to provide a significant reduction in the temperature component $T_{n,E}$. Moreover, at low altitudes atmospheric noise will also reach the antenna through the side lobes, and the proposed system will help to attenuate it.

The technology of radio receivers has now developed to the point where major problems are arising from radio interference coming from industrial sources and from distant radio transmitters working at the wavelength of the receiver. Waves in the centimeter range are not fully damped out through scattering by inhomogeneities in the atmosphere and ionosphere even within distances of hundreds or thousands of kilometers, and they contribute appreciable interference in observational work.

One possible way to combat interference would be to reduce the size of the side and back lobes of the reception pattern of an antenna system. In fact, the problem of reducing the radiating effect of the earth is essentially an equivalent one. The system proposed here serves to diminish the size of the side lobes relative to the systems customarily used.

We turn now to a description of the proposed antenna system; although it retains the basic elements of a parabolic antenna, it provides a reduction in the size of the side lobes:

a) Consider Fig. 1. The treatment will first proceed in the geometric—optics approximation. Considerations of this type become increasingly valid for short wavelengths and large reflector diameters.

In order to prevent interference from reaching the feed, the paraboloid could be covered by a metal screen $S_1-S_1-S_1$. Let the screen first take the form of a spherically cylindrical screen with radius R_1, as shown in Fig. 1. This system would evidently be free of interference, but it could not receive signals either.

b) Now let us shorten the radius R_1 and install a screen of radius $R_2 < R_1$, as shown in Fig. 2. The screen still shields the feed located at the focus F from reception, substantially reducing the energy received from directions outside the paraboloid.

If $R_2 = kR_1$, it is clear that the screen will cast a shadow of area $A = k^2 A_1$, where A_1 is the total area of the paraboloid aperture.

In practice we may, for example, take $k = 0.2$ or 0.1, so that the vignetted area will be only a few percent of the total area and the decrease in efficiency will be small.

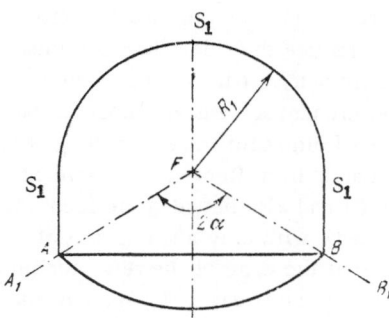

Fig. 1. Design for a "deaf" screen. F) Focus of paraboloid; FA_1, FB_1) rays bounding the aperture angle 2α of the paraboloid.

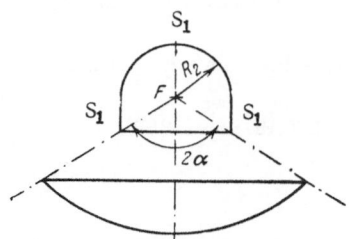

Fig. 2. Design for a feed screen.

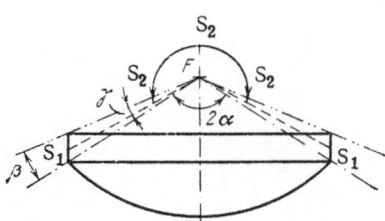

Fig. 3. Design for shielding of the feed and the reflector edge.

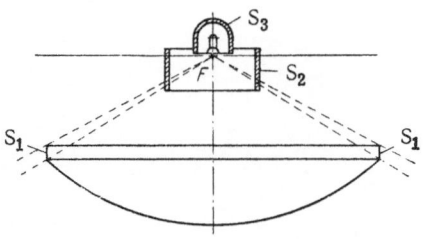

Fig. 4. Design for a ring feed screen.

c. On the basis of design considerations for mounting the screen (Fig. 3), we could put part of the screen at the periphery of the reflector and transfer part of it, with a smaller radius, to the center, so that the screen would be divided into a ring S_1 mounted at the edge of the reflector and a component S_2 placed closer to the focus. To diminish the effects of diffraction at the edges, there is nothing to prevent us from achieving more shielding by replacing the aperture angle $\xi = 2\alpha + 2\beta$ of the screen S_2, the value that would follow from the condition that one screen be supplemented by the other, by the somewhat smaller angle $\xi = 2\alpha + 2\beta - 2\gamma$. Here 2γ is the "overlap" angle of the two screens, that is, the angle within which screens S_1 and S_2 would simultaneously provide geometric shielding against waves approaching the focus.

d. Now let us consider the problem of further reducing the area vignetted from the reflector.

In several cases we can evidently open up the screen S_2 on top, or subdivide it into two parts, obtaining the design shown in Fig. 4.

Here the screen is in fact divided into three parts: $S_1 + S_2 + S_3$. Perhaps, in some cases, one could manage without the screen S_3. The question of the dimensions of the screens, their mutual overlapping, and the structural design should be investigated and analyzed in each individual case.

Let us examine the harmful effects of diffraction at the edges of the screens. The values of the diffraction angles can be very appreciable, and they will restrict the opportunities for bringing the screen S_2 (Fig. 3) or the screens S_2 and S_3 (Fig. 4) close to the feed.

To diminish diffraction effects, which are relatively weak at short wavelengths, and thereby halt deterioration of the operating quality of the system, the following measures may be recommended:

a. Select the best location for the screens, and in particular make a good choice for their distances from the focus (R_2 in Fig. 2).

b. Arrange the screens so that they overlap. Make a judicious choice for the angle γ (Fig. 3) and for the overlap angle Θ of screens S_2 and S_3 (Fig. 4).

c. Cover the edges of the screens with an absorbing layer or fabricate the edges of the screens of network with a variable mesh, increasing outward in area.

Each of these measures, and all in combination, should be specially studied and applied differently in every particular case.

The screens could be made of sheet metal or of network furnishing adequate reflection of radio waves in the operating range.

To handle particular problems of screen design, one should note that waves can reach the feed not only directly but also after double or multiple reflection. For instance, Fig. 5 illustrates a design in which this situation occurs. The dimensions, shape, and orientation of the screens should be determined specifically for each case. Figure 6 shows one successful design for the screen S_2.

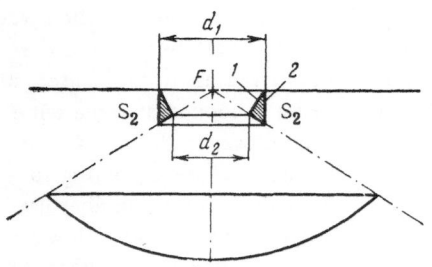

Fig. 5. Design for a nonferrous feed screen. 1) Truncated cone; 2) cylinder; d_1, d_2) diameters of clear apertures.

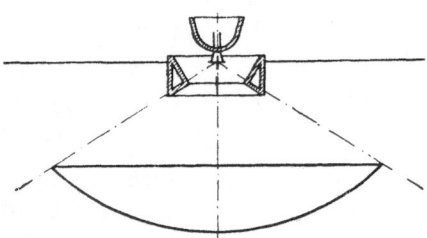

Fig. 6. Design for a double feed screen.

The distinctive property of the reflector antenna proposed here is that in order to reduce the side lobes, diminish the effect of radio emission from terrestrial sources, and enhance the resistance to interference, screens are mounted on the antenna to shield the feed from radio waves reaching it from directions outside the parabolic reflector. The screens are placed directly at the feed and also between the feed and reflector (but closer to the feed), with only a small part of the shielding located directly at the edge of the reflector (in principle, this part could even be omitted). The screens may be of metal (sheet or mesh), with or without partial covering by a layer that absorbs radio waves. The screens may partially overlap each other, and may overlap the reflector itself somewhat in feed shielding angle.

Literature Cited

1. J. G. Bolton, "Radio telescopes," in: Telescopes, G. P. Kuiper and B. M. Middlehurst (eds.), Univ. Chicago Press (1960), p.176.
2. P. D. Kalachev and I. A. Salomonovich, Radiotekhn. i Elektron. 6(3) (1961).
3. B. I. Klass, Aviation Week 78(5): 80 (1963).

PROBLEMS IN CONSTRUCTING A
HIGH-RESOLUTION PARABOLIC ANTENNA

P. D. Kalachev

Introduction

The further successful development of radio astronomy and space radio communication requires the construction of large antennas, especially dish antennas of considerable size and high accuracy. In radio astronomy such antennas, supplied with radio receiving apparatus of high sensitivity, are called radio telescopes.

The efficiency of a radio telescope is determined by two principal characteristics: resolution and sensitivity. The resolution of a radio telescope is governed by the extent of the reflecting surface of its antenna in two (or at least in one) directions and the wavelength of the received radio emission. Sensitivity is determined by the effective area of the reflecting surface of the antenna and by the sensitivity of the radio receiving apparatus [1,2]. An effort to obtain the optimal combination of these two characteristics, taking into account the specific purpose of the apparatus and its cost, has resulted in the appearance of a great variety of designs of radio-telescope antenna systems. We will list some of them.

1. Dish antennas in the form of a "classical" paraboloid of revolution with full or restricted turning of the reflector about two mutually perpendicular axes.
2. Plane cophased antennas with turning about the vertical and horizontal axes.
3. Cross-shaped antennas of the "Mills Cross" type with wire reflectors, oriented in East—West and North—South directions.
4. Fixed spherical "ground" dishes with a movable small reradiating reflector.
5. A variable profile antenna of the type developed by Khaikin and Kaidanovskii [3].
6. Multielement antennas with the reflecting elements arranged in the form of a cross or in a closed circle (in layout).
7. Antenna systems operating on the interferometer principle, consisting of two or more individual reflectors separated by some distance.

The antenna systems mentioned above can be used in layouts known as antennas with aperture synthesis, which, for attaining a high resolution, require special processing of the received signals.

Among the types of antenna systems mentioned, the fully steerable parabolic reflector possesses a number of advantages: great universality, broad range capability, possibilities for simultaneous work at several wavelengths, possibilities of tracking the observed object and rapid change of direction, low noise temperature, and others.

A serious shortcoming of a fully steerable parabolic reflector of considerable size* is the great difficulty involved in its design and construction. Regarding a comparison of parabolic antennas and equivalent antennas of other types from the cost point of view, it is difficult to draw definite conclusions concerning the economic advantages of the different types, although the latter are developed to a considerable degree with cost considerations in mind.

---

*A reflector with a diameter greater than 20 m may be said to be of considerable size.

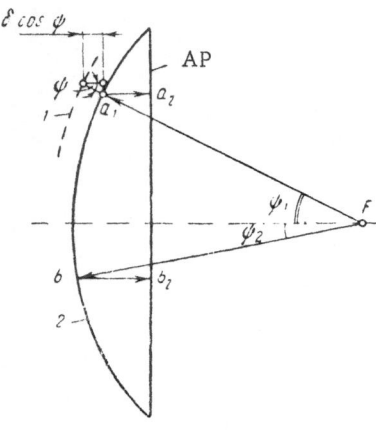

Fig. 1. Diagram of a paraboloid surface with a distorted sector. 1) Distorted sector of a parabolic surface; 2) ideally perfect parabolic surface of reflector.

If a comparison is made on the basis of the size of the reflecting surface of antennas for work with long waves (such as those in the meter range), the economic advantages of cross-shaped antennas with wire reflectors is indisputable.

Taking the above into account, it can be said that the possibilities of parabolic antennas have not yet been exhausted. We feel it desirable to make efforts to find such designs which will ensure the possibility of constructing a large parabolic reflector of high accuracy and which is sufficiently rigid for maintaining a true parabolic reflecting surface whatever spatial position the reflector assumes.

Requirements on the Accuracy of the Reflecting Surface of a Parabolic Reflector

As is well known, the theoretically perfect reflecting surface of the reflector of a transmitting antenna should ensure a cophased character of the field in its aperture. The latter is achieved by the equality of the total length of the path of any beam from the exciter situated at the focus of the reflector to the reflecting surface and from the reflecting surface to the aperture plane of the reflector (Fig. 1), that is,

$$Fa_1 + a_1 a_2 = Fb_1 + b_1 b_2.$$

In a simple single-dish system, the surface which has the capability of creating the desired cophased character of the field in the aperture with radiation from a single point is usually a paraboloid of revolution. In two- or multidish systems [4], the reflecting surfaces may not be paraboloids of revolution.

It is necessary to create a cophased field in the aperture in order to ensure the maximum possible antenna gain. Since in actuality it is impossible to produce an absolutely precise reflecting surface and also impossible to maintain it unmodified during the course of its use, the problem arises of the admissible errors, that is, the admissible deviations from a theoretical (especially parabolic) reflecting surface.

Departures from a parabolic form of the reflecting surface can be of two kinds: local displacements of parts of the surface along the normal — depressions and bulges causing an increase or decrease of the total path of the beam reflected at the site of the bulge or depression, and local changes of the slope of the plane tangent to the paraboloid from the theoretical value, which means a change of the direction of the beam reflected from a particular point of the reflector. This results in a blurring of the focal spot in the focal plane.

According to a number of investigators [4,5,6], the admissible deviation from a cophased field in the aperture can be accepted to be a value equal to $\pm \pi/4$ [5].*

The total length of the path of the beam from the focus F to the aperture plane AP of the reflector for an ideal paraboloid is

$$S_0 = Fa_1 + a_1 a_2.$$

For a distorted sector of the paraboloid the change of the length of the path is

$$\Delta S = \delta (1 + \cos \psi).$$

---

*A radio telescope with a parabolic reflector 37 m in diameter was constructed recently in the United States for work in the centimeter range to $\lambda = 1$ cm. The reflecting surface of the reflector of this radio telescope is made with an accuracy to ± 0.25 mm, which is indicative of the greatly increasing demands being imposed on the accuracy of a reflecting surface [7].

Since the phase change $\Delta \varphi$ of a variable electromagnetic field of a wave is proportional to the length of the path traveled by the wave, then

$$\frac{\Delta S}{\lambda} = \frac{\Delta \varphi}{2\pi} \; ; \quad \text{hence} \quad \Delta S = \frac{\Delta \varphi}{2\pi} \lambda$$

or

$$\delta \, (1 + \cos \psi) = \lambda \frac{\Delta \varphi}{2\pi} \, ,$$

where λ is wavelength. Hence,

$$\delta = \lambda \; \frac{\Delta \varphi}{2\pi \, (1 + \cos \psi)} \cdot \tag{1}$$

Assuming the maximum admissible phase shift to be $\Delta \varphi = \pi/4$, we have

$$\delta = \frac{\lambda}{8 \, (1 + \cos \psi)} \cdot \tag{2}$$

At the center of the reflector $\psi = 0$, so that

$$\delta_{\psi=0} = \frac{\lambda}{16} \, . \tag{3}$$

Formulas (1) and (2) show that the admissible error is minimal in the central part of the reflector and increases toward its edges.

In the case of short-focus parabolic reflectors ($2\psi \approx 180°$), the value of the admissible error at the edges increases by a factor of two, whereas for long-focus reflectors the increase is insignificant. For example, for the parabolic reflector of the FIAN-RT-22 radio telescope [1] with an aperture angle $2\psi = 120°$, the admissible error of the surface at the edges of the reflector is

$$\delta = \frac{\lambda}{8 \, (1 + \cos 60°)} = \frac{\lambda}{12} \cdot$$

In the case of the 25-m reflector of the Bonn radio telescope, having an aperture angle $2\psi \approx 140°$, the admissible error at the edges of the reflector is

$$\delta = \frac{\lambda}{8 \, (1 + \cos 70°)} \approx \frac{\lambda}{10.7} \cdot$$

In comparison with the ideal surface of a reflector, the energy losses in scattering due to the "roughness" of the reflecting surface with errors not exceeding $\lambda/16$ will not be greater than 5% [5]. In the case of errors not exceeding $\lambda/10$, the losses will be about 20%. A further increase of the errors of the reflecting surface of the reflector leads to a rapid increase of the energy loss. For example, in the case of errors $\lambda/5$, the losses already are more than 50%.

Deformations of the Reflector. Formulation of the Problem

Having reviewed the requirements imposed on the accuracy of the reflecting surface of the radio telescope antenna, we shall consider the factors responsible for deviation from the theoretical form of this surface. These factors can be classified into two groups:

1. Engineering errors δ_{eng}, inevitable in the construction process. They have a constant character and do not change thereafter.
2. Elastic deformations (variable in value and direction), dependent on environmental conditions. In turn, elastic deformations can be divided into: a) elastic deformations δ_{ed}^{wt}, caused by the load imparted by the weight of the apparatus itself; b) elastic deformations δ_{ed}^{wd}, caused by

wind loads; c) temperature (reversible) deformations δ_t, arising due to the uneven heating of the elements of the framework and surface components of the reflector.

In order to determine the value of the total error, it is necessary to clarify the problem of the method for their addition.

The elastic deformations from the effect of weight and wind are distributed over the reflector and are directed in accordance with the loads primarily in such a way that they can be added algebraically, that is,

$$\delta_{ed} = \delta_{ed}^{wt} + \delta_{ed}^{wd}$$

Temperature deformations are not related directly either to elastic deformations or to the engineering errors in construction. The same can be said of the relationship between engineering errors and elastic deformations. However, something can be said of an indirect relationship between engineering errors and temperature and elastic deformations. The fact is that all the errors mentioned increase at the edges of the reflector.

Engineering errors increase toward the edges due to two factors: a) there is an increase of pattern errors with an increase of distance from the axis of rotation (the geometrical axis of the paraboloid of revolution) because the measurement errors increase; b) there is an increase of the error (although only insignificantly) of measuring the position of the reflecting surface (covering of the reflector) in relation to the pattern as a result of increase of the sensitivity of the pattern on the large arm to small loads during measurements (deformations of the pattern). In the case of a two-sided pattern, when its size is equal to the diameter of the reflector, these factors are virtually absent.

Elastic deformations also increase in the direction of the edges of the reflector [see formula (8)]. It is probable that temperature deformations also increase in the direction of the edges of the reflector, if for no other reason than that with equal temperature gradients the temperature deformations at the edges will be greater due to the lesser rigidity of the reflector framework at the edges.

With respect to the maximum possible errors — departures of the reflecting surface from the stipulated form, these are determined as the arithmetical sum

$$\delta_{max} = \delta_{eng} + \delta_{ed} + \delta_t. \tag{4}$$

However, the maximum possible errors do not determine the quality of the reflecting surface because they are relatively rare. We feel that the most probable errors are the total errors determined as the mean square sums

$$\delta_\Sigma = \sqrt{\delta_{eng}^2 + \delta_{ed}^2 + \delta_t^2}. \tag{4'}$$

With respect to the terms mentioned, the first two have been studied most thoroughly; the least is known about the third.

Engineering errors are determined directly in the course of construction and adjustment of the reflector. Elastic deformations are detected primarily from static computations and subsequent investigation of the reflector. During construction of the 22-m reflector of the RT-22 radio telescope, no basic difficulties in increasing the accuracy of construction of this reflector, completed with a maximum error of ±0.8 mm, were encountered.

With an increase of the diameter of the reflector, the construction errors naturally will increase. If the installation and adjustment of the covering of the reflector are done using an engineering control pivoting pattern, the accuracy of construction of the reflector will be determined by the accuracy of construction of the pattern itself and the accuracy of its installation. Deviations of the working surface from the pattern, measured by surface-type or through-type thickness gages, are to all intents and purposes not dependent on the size of the reflector, provided the required measures are taken to ensure stable operating conditions (perfect revolution with the pattern, constancy of temperature, absence of meteorological interference, etc.). It can be assumed, apparently, that the possible errors in constructing the engineering control pattern increase proportionally to the linear dimensions of the reflector, i.e., that

$$\delta_{pat} = \pm kD_{refl} ,$$

(5)

where D_{refl} is the diameter of the reflector.

It was established from repeated careful adjustments of the pattern prepared for construction of the 22-m FIAN radio telescope that the maximum errors in its preparation (deviations from the stipulated parabola) did not exceed ± 0.5 mm.

Substituting the numerical values of the error $\delta_{pat} = \pm 0.5$ mm and the diameter of the reflector $D_{refl} = 22$ m into relation (5), we obtain k = 0.00002 and, therefore,

$$\delta_{pat} \cong \pm 0.00002 D_{refl}.$$

(6)

For the pattern required, for example, in constructing a reflector three times the size of the RT-22, the anticipated maximum error in preparation will be

$$\delta_{pat} \approx \pm 0.00002 \cdot 66.000 \approx \pm 1.5 \text{ mm.}$$

In checking the covering of the RT-22 reflector, it was found that the maximum deviations from the pattern were ± 0.6 mm. As already mentioned, for a 66-m reflector these errors will remain virtually the same. This consideration, among others, is an advantage of constructing a reflector using a pattern. Apparently, it is most probable that the errors in preparing the pattern δ_{pat}, and the errors in deviation from the pattern $\delta_{dev pat}$ detected in checking should be added geometrically, and that the total engineering error can be determined using the formula

$$\delta_{eng} = \pm \sqrt{\delta^2_{pat} + \delta^2_{dev pat}} \quad \delta_{eng} \approx \pm \sqrt{1.5^2 + 0.6^2} \approx \pm 1.6 \text{ mm.}$$

(7)

As already mentioned, the elastic deformations of the reflector caused by loads from the weight of the reflector can be determined by static computations of the reflector. The multiple connections of the reflector-suspension system, even when there is symmetry of design, makes it necessary to solve a system of canonical equations with many unknowns. In order to obtain more precise and especially, more reliable data from static computations, it is required that the theoretical computations be accompanied by experimental computations using the results of measurements of strains and stresses obtained using models.

At present there are no known experimental data on determination of the flexure of a reflector with effective loads under operating conditions. The only mention in the literature is that during preparation of the reflector of the 25-m radio telescope of Bonn University, static tests were made for the purpose of checking the deformations of the reflector obtained as a result of static computations [8]. The tests were made in the workshops of the manufacturing plant. The results of the tests revealed satisfactory agreement with computed data, which is evidence of the possibility of obtaining quite reliable data on elastic deformations by computation methods alone.

The elastic transverse deformations of the beam caused by a transversely distributed load can be expressed, as is well known (from texts on the resistance of materials), by the following formula:

$$\delta^{wt}_{ed} = \frac{ql^4}{c_1 EI} + \mu \frac{Ql}{GF} ,$$

(8)

where the first term is the flexure of the beam from a distributed load q [kg/cm] (for example, structural weight) and is governed by the effect of lateral flexure from the load q; the second term is the flexure of the beam from the transverse force Q, which is a function of the transverse load.

In formula (8), l [cm] is the value of the span (or cantilever) of the beam; E and G [kg/cm^2] are the moduli of elasticity of the material of the structure for the normal and shearing stresses, respectively; I[cm^4] is the moment of inertia of the cross section of the beam; F[cm^2] is the area of the cross section of the beam; μ is the shape factor for the cross section of the beam; c_1 is a factor dependent on the end attachments of the beam. Since the elastic deformations of the reflector are determined by the transverse deformations of the main supporting elements of the framework — the radial beams, the flexure computed using formula (8) is the principal component of the elastic deformations δ_{ed} of the reflector.

The value of the second term is dependent on the ratio of the height of the cross section (construction height) to the length of the span (cantilever). If the ratio mentioned is less than 1:5, the second term need not be taken into account, because it will be less than 10% of the total flexure.

Since it is the flanges of the beams which for the most part absorb flexure, for ordinary structures the moment of inertia can be determined as the product of the square of the cross section of the beam and the area of the cross section of the flanges

$$I = kH^2 F_{fl}. \tag{8'}$$

The sectional area of the flanges is proportional to the square of the length of the flange

$$F_{fl} = k_{fl} l^2; \tag{9}$$

the sectional height is proportional to the length of the beam

$$H = k_H l. \tag{10}$$

The weight of the beam is

$$G_{wt} = k\Sigma l_{web} F_{fl.web}, \tag{10'}$$

where l_{web} is the length of an element of the web or flange, $F_{fl,web}$ is the sectional area of an element of the web or flange. The distributed load (running weight of the beam) is equal to

$$q = \frac{G_{wt}}{l}. \tag{10''}$$

The total length of the elements of the web and flanges when H/l = const is proportional to the length of the beam

$$\Sigma l_{web} = k_{web} l. \tag{11}$$

The transverse force is proportional to the weight of the beam

$$Q = k_Q G_{wt} = k_Q (k\Sigma l_{web} F_{fl.web}). \tag{12}$$

Substituting the corresponding values from (8'), (9), (10), (10'), (10''), and (12) into (8), we obtain

$$\delta_{ed}^{wt} = \frac{k_2 l^3 l^3}{cEk_1 l^4} + \frac{k_Q l^3 l}{Gk_{fl} l^2} \mu = kl^2, \tag{13}$$

where

$$k = \frac{k_2}{cEk_1} + \mu \frac{k_Q}{Gk_{fl}}.$$

Expression (13) shows that the elastic deformations of the supporting elements of the framework of the reflector are proportional to the square of their lengths, which in turn are proportional to the diameter of the reflector

$$l = c_{web} D_{refl}.$$

Therefore,

$$\delta_{ed}^{wt} = k_{el} D_{refl}^2. \tag{14}$$

Determination of the values of elastic deformations of the reflector due to the weight of the structure δ_{ed}^{wt}, both theoretically (by means of static computations) and experimentally, presents no basic difficulties, although it involves unwieldy computations.

The following considerations can be mentioned with respect to the elastic deformations of the reflector from wind loads:

1. The loads may be in any direction.
2. The distribution of loads over the structure of the reflector is extremely complex and in general form theoretical computations cannot be made; by exposure of the model in a wind tunnel, it is possible to obtain a load curve more or less corresponding to the true curve, but this is dependent on the quality of modeling.
3. Whereas, for an isolated paraboloid, the wind load curve would have one axis of symmetry (along the flow velocity), a more complex load curve will be caused by the presence of large "appendages" (shafts, supporting the exciter; supporting structure; rotating sectors, etc.) distorting the symmetry of the flow. This circumstance considerably complicates both the theoretical computation of elastic deformations and their experimental checking.

During wind-tunnel exposure of a model of a reflector with a circular deflector [9], the aerodynamic coefficient of head resistance is

$$c_x = 1.9.$$

The wind load per square meter of the reflector will be

$$q_v = c_x \frac{v^2}{16} = 1.9 \frac{v^2}{16} = 0.12 \ v^2 \left[\frac{\text{kg}}{\text{m}^2} \right]. \tag{15}$$

In computations of the elastic deformations of a reflector, it is undesirable to use a computed wind velocity of greater than 12 m/sec because, in the middle latitudes of the USSR, the wind velocity is 12 m/sec or less during 95% of the entire year.

At such a wind velocity, the load per square meter of the reflector will be

$$q_v = 0.12 \cdot 12^2 \approx 17 \frac{\text{kg}}{\text{m}^2}.$$

Naturally, computations for stability and overturning of the entire antenna apparatus should be made for winds of hurricane force (40-50 m/sec).

For a rigid reflector, designed for operation on short waves in the centimeter range ($\lambda = 1$-3 cm), the load from the weight of the structure per 1 m² of the aperture area is an order of magnitude greater than the operational wind loads. For example, the load from the structural weight of the FIAN RT-22 radio telescope 22-m reflector is 170 kg/m². The load from the structural weight of the 42-m reflector of a radio telescope constructed in the United States, designed for operation in the centimeter wavelength range (up to $\lambda = 3$ cm), according to preliminary data, is about 220 kg/cm².

Assuming in the first approximation that there is a uniform distribution of wind loads over the surface of the reflector, it can be assumed that the increase of elastic deformations of the reflector from wind loads (δ_{de}^{wd}) also will be proportional to the square of the diameter of the reflector, similar to the elastic deformations from structural weight.

For a more precise determination of deformations of the reflector from wind loads, it is necessary to have more precise information on the values of the aerodynamic moments and, therefore, on the distribution of wind loads over the surface of the reflector for different angles of attack. This information should be studied not only by exposure of large drained models in a wind tunnel, with detailed reproduction of the appendages (framework of the reflector, large rotating sections, suspension structure of the exciter, etc.), but also by field measurements of the aerodynamic moments on real radio telescopes.

With respect to temperature deformations, the lack of experimental data on the temperature gradients in the elements of the framework and covering of the reflector makes it impossible to determine them with any reliability.

The author knows of no published results of theoretical analysis of this problem, with the exception of an investigation of the special case of uniform heating or cooling of the reflector [8, 10]. However, this case is less interesting because, in the case of uniform heating (cooling) of a reflector, its efficiency should not decrease. The fact is that in the case of a uniform temperature change of the elements of the reflector there is only a change of its parameter (focal length) and there is no disruption of the cophased character of the field in the aperture or coincidence of the exciter with the focus of the reflector, provided all the elements of the reflector and the structure supporting the exciter are made of materials with identical linear coefficients of expansion. It can be said that the degree of nonuniformity of heating of the components of the reflector is dependent on the incidence of direct solar radiation on them and also on wind velocity and direction. At nighttime, and on overcast days, the nonuniformity of heating should be minimal.* Study of deformations from nonuniform heating makes it possible to adopt sound solutions or measures for decreasing or compensating these deformations.

The full solution of the problem of the deformations of parabolic reflectors thus requires further theoretical and experimental investigations.

The purpose of this study is an attempt to solve the less general problem of constructing a parabolic reflector having quite small deformation under the influence of the principal factor responsible for deformation, i.e., the influence of loads from structural weight. We shall solve this problem by rational selection of design variants for the framework of the reflector and its suspension to the supporting and steering apparatus.

Design Variants for the Reflector and Its Suspension

If we consider the design variants for the supporting framework of a reflector and its suspension to the supporting-steering apparatus in all presently existing radio telescopes, it is easy to note the following. First, the supporting framework of the reflector in most cases is radial-symmetric. Exceptions to this rule are three parabolic antennas known to the author which are constructed on the principle of parallel main supporting elements: German — "Great Würzburg" [11], the truncated "V-3" parabolic antenna of the FIAN [12], and the 25-m reflector of the radio telescope at Dwingeloo (Netherlands) [13]. The selection of a radial-symmetric design is entirely sound both from the structural and technical (use of symmetry in designing structural components, identity of elements and units, etc.) points of view. Second, in most variants the method for suspending the reflector from the supporting-steering structure is close in principle to a two-support beam. There are exceptions in this case as well: the Australian radio telescope with a 64-m reflector [14] and the FIAN radio telescope with a 22-m reflector [1]. Figure 2 shows diagrams of the suspensions of a reflector to the supporting steering structure.

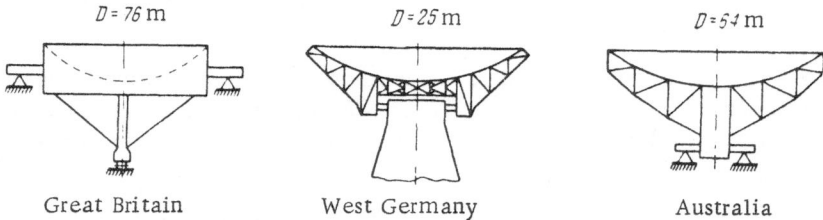

Fig. 2. Diagram of the reflector suspension on the supporting-steering structure.

---

*A recently published study [15] discusses the influence of solar heating on the form of the parabolic dish of a radio telescope.

The suspension of the reflector of the 76-m radio telescope at Jodrell Bank (Great Britain) [11] is similar to a three-support beam for which the central support is elastic. The suspension of the 25-m Bonn (West Germany) radio telescope is similar to a two-support beam. The suspension of the reflector of the Australian radio telescope to all intents and purposes can be considered radial-symmetric. The reflector is suspended on the base of the central support of the framework; it has a diameter of approximately 5 m, which is 7-8% of the reflector diameter. In comparison with the framework of the reflector, the central support can be considered absolutely rigid.

If the reflector were fixed and its geometrical axis directed to the zenith, the problems of ensuring only small elastic deformations would not be substantial even for a reflector with a diameter of several hundred meters. Since the number of support points for a fixed reflector can be as great as desired, the size of the cantilevers and spans, and therefore the flexural deformations, can be reduced to a minimum. However, operational requirements make it necessary that the reflector be directed to any point of the celestial sphere and, therefore, the design of the suspension of the reflector to the supporting-steering structure should ensure that the reflector can turn about two mutually perpendicular axes.

In the case of an azimuthal setting, these axes are oriented vertically and horizontally. Ensuring turning about the vertical axis exerts no appreciable influence on the reflector suspension. However, the requirement for turning about the horizontal axis imposes at least two important conditions on the suspension design: first, the supporting bearings must be situated on a single straight line (horizontal axis) and second, there must be a free space between the supporting bearings (or vice versa, on the outer side of the bearings), ensuring room for the reflector to turn about this axis. Taking the above into account, designers have been forced to suspend the reflector using the two-support beam principle.

The framework of a reflector, designed using a radial-symmetric arrangement of components, has a maximum rigidity when there is symmetrical loading and minimum rigidity when there is skew-symmetric loading, since, in the first case, the circular elements of the framework increase the rigidity of the reflector and, in the second case, they are only connecting components of the framework, facilitating redistribution of the loads.

In order to clarify the influence of the supports (suspension) of the reflector framework on its deformation, we will consider the framework of a reflector with a radial-symmetric design (Fig. 3).

If two supporting journals are attached directly to the supporting framework at the points KK, even when the reflector is in a horizontal position during radial-symmetric loading by the structural weight, the movements of the points of the framework (as a result of elastic deformations) lying on a circle of the same radius will be different: the farther the considered point is from the supporting journals (along the circle), the greater is its movement. If the supporting journals are attached directly onto the central support of the supporting framework, at the points TT, in a case when the central support is of small diameter in comparison with the diameter of the reflector (Fig. 2, Australian model), the movements of the points of the framework lying on a circle of the same radius (at symmetric points relative to the radial elements of the framework) will be virtually identical and the supporting attachment of the reflector can be considered radial-symmetric.

In the case of a two-support suspension of a noncircular reflector, there is an optimal relation between the suspension base KK and the diameter of the reflector D_{refl}. For a narrow "break" (a parabolic reflector with cutoff segments), denoted in Fig. 3 by the numbers 3', 4', 7', and 8', the following relation is correct:

$$KK \approx 0.55 D_{refl}. \tag{16}$$

As shown by computations made recently at the radio astronomy laboratory of the FIAN, this relation does not apply.*

Computations reveal that in the case of ordinary two-support suspension, the maximum elastic deformations from structural weight at the edge of a 22-m reflector in a horizontal position are approximately 5 mm.

---

*See "Elastic Deformations of the 22-Meter Parabolic Reflector of the RT–22 Radio Telescope of the Lebedev Institute (FIAN) due to gravity loading of the Structure," by P. D. Kalachev, this volume, p. 143.

Decrease of the deformations mentioned by a simple increase of the cross sections of the supporting elements cannot be achieved, due to an increase of loads from structural weight. We will demonstrate that an appreciable decrease of deformations without a change of the cross sections of the supporting elements of the structure can be attained by a change in the method of suspending the reflector.

This is illustrated by the fact that a changeover in the RT-22 from a two- to a four-support system led to a decrease of maximum deformations (from symmetrical loads) on the edge of the reflector from 5 to 0.78 mm, that is, by more than 6 times.

In addition, a four-support suspension of the reflector ensures a more rigid suspension of the exciter.

New Method of Reflector Suspension

It is easy to see that such a method should be a multisupport design of suspension of a reflector with a radial-symmetric arrangement of supports. Figure 3 shows the plan for a reflector-supporting system with eight supports. It should be noted that the fulcrums 1, 2, . . . , 8 are not situated on the radial supporting elements of the framework, but in the intervals between them, on the chord members. In this case, the number of supports required for maintaining the radial symmetry of suspension of the reflector is half as great as when the fulcrums are on the radial supporting elements. The figure shows that each radial supporting element is arranged identically relative to the fulcrums.

The proposed method for solution of the problem of small deformation of the reflector can be summarized as follows.

1. It is necessary to determine both the minimum number of fulcrums necessary to ensure the required rigidity for a reflector of a particular size and the admissible elastic deformations.
2. It is necessary to find a design variant for suspension of the reflector which will ensure the effectiveness of all the fulcrums as if they were on an absolutely rigid body, i.e., the relative arrangement of the fulcrums will remain unchanged regardless of the position of the reflector in space. The first part of the problem is an ordinary problem in structural mechanics and due to inadequate space will not be considered in this study. However, we are concerned primarily with the second part of the problem, which is basic.

It was mentioned earlier that in order to ensure the possibility of turning of the reflector an elevation angle, i.e., about its horizontal axis, the supporting bearings of the reflector must be situated on a straight line, which is also the horizontal axis, thus ensuring free movement of the reflector during its rotation.

Usually the reflector is suspended using two supporting bearings situated on the base $B = (0.4-0.5)D_{refl}$. Therefore, whatever the number of reflector fulcrums, in the last analysis they must be reduced to the two supporting journals. The most graphic and convenient design variant solving this problem is a design having a number of fulcrums expressed by the equation

$$z = n2^m + k, \qquad (17)$$

where z is the number of fulcrums, n is the number of concentric circles on which the fulcrums are situated, 2^m is the number of fulcrums in each circle, and k is the number of central fulcrums, with n = 1, 2, 3, . . . , m = 2, 3, 4, . . . , k = 0 or 1.

For example, the design variant for the 22-m reflector of the FIAN RT-22 radio telescope involves use of four symmetrically arranged supports (n = 1, m = 2, and k = 0). Figure 3 shows the rotating assemblies, combining a four-support suspension of the reflector with two supporting journals corresponding to the two supporting bearings.

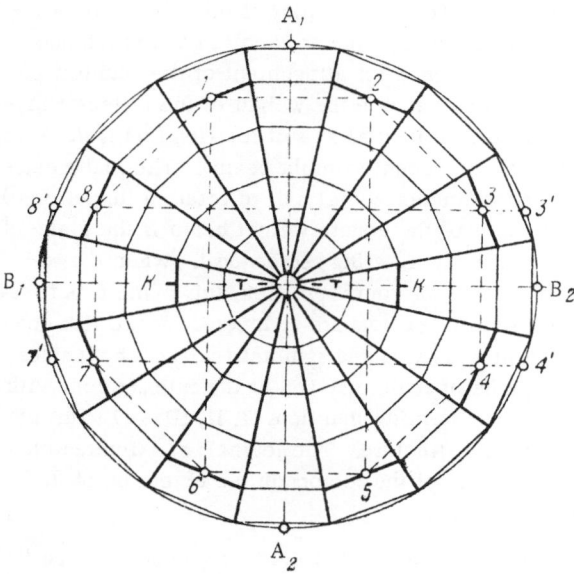

Fig. 3. Plan of reflector supporting system.

In the 22-m RT-22 reflector, the four fulcrums are situated on the upper flat surfaces (crossbars) of the rotating assemblies with a radial-symmetric arrangement among the 16 supporting radial elements of the reflector framework and this ensures its greater rigidity. As noted above, the computed flexures at the edge of the reflector resulting from symmetrical loads do not exceed 0.8 mm. The diameter of a circle passing through the four fulcrums is 50% of the reflector diameter. In the case of skew-symmetric loading, the computed elastic deformations at the edge of the reflector do not exceed 2.5 mm.

Using as a point of departure the values of the admissible elastic deformations and the experience in constructing the RT-22, there is sufficient basis for the assumption that the framework parts of a parabolic reflector having a relative construction height $H/l = 0.2$-0.3 can be equal to 3-5 and 12-16 m, respectively, for the cantilever l_c and spin l_s.

On this basis, and using relation (17), without recourse to the general theory of the dependence of the number of supports on the diameter of the reflector, for a given value of elastic deformation we can determine the required number of supports for different diameters. For example, for a reflector with a diameter $D_{refl} = 35$ m, the required number of fulcrums will be determined by the values n = 1, m = 3, k = 1, or

$$z = 1 \cdot 2^3 + 1 = 9.$$

i.e., in each diametral section of the framework of the reflector passing through the fulcrums there will be three supports (see Fig. 3).

For a reflector with the diameter $D_{refl} = 45$ m, assuming n = 2, m = 3, and k = 0, we have

$$z = 2 \cdot 2^3 + 0 = 16.$$

In this case, the fulcrums are situated in two circular sections of the framework with eight points in each. The diameter of the outer circle is 40 m and the diameter of the inner circle is 14 m. In each diametral section of the reflector framework passing through the fulcrums there are four supports.

As an example, we will consider in greater detail the design of a multisupport suspension of a reflector with the diameter $D_{refl} = 66$ m. First, reasoning much the same as before, we conclude that 17 fulcrums are required, arranged radial-symmetrically relative to the reflector framework. In this case, the 17 fulcrums were as follows: number of concentric circles, n = 2; number of fulcrums in each circle, $2^m = 2^3 = 8$; number of central fulcrums, k = 1; total number of fulcrums

$$z = n2^m + k = 2 \cdot 2^3 + 1 = 17.$$

Now we will proceed to the main problem: consideration of a design variant in which all 17 supports are effective (regardless of the position of the reflector in space), as if all the supports were absolutely rigid.

There are three principal parts of the considered design variant for suspension of the reflector: the rotating sectors (Fig. 4), the intermediate eight-support structure (Fig. 5), and an eight-shaft pyramid with a central support (Fig. 6). As shown by Fig. 4, the unit formed by the rotating sectors contains two driving toothed or cog sectors (1) connected by a pipe (2) which is the horizontal axis of rotation of the reflector, and a system of parallel or semidiagonal shafts, forming a rigid spatial system capable of absorbing the flexural moments

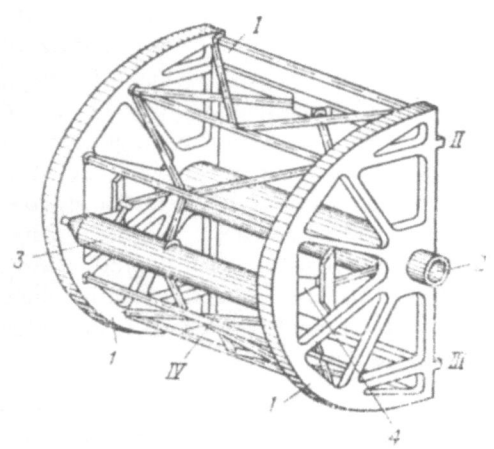

Fig. 4. Rotating sectors unit.

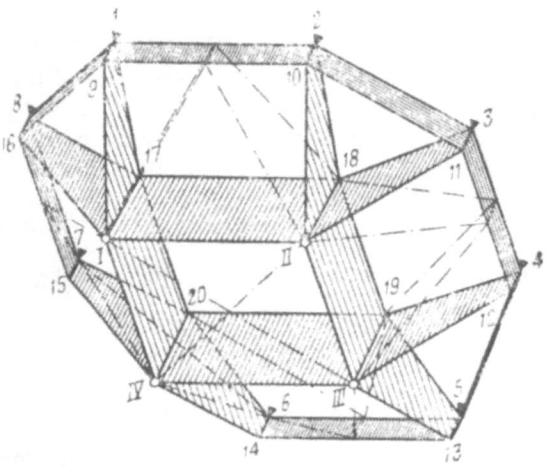

Fig. 5. Intermediate eight-support structure.

and torques. The lower points of the rims of the sectors are connected to the horizontal counterbalance beam (3), designed for attachment of the counterweight weights which balance the weight of the reflector in relation to the horizontal axis of rotation [(4) is the lever for moving the counterweight beam]. The plane passing through the horizontal axis of rotation (upper pipe) and the axis of the counterweight beam is the plane of symmetry of the rotating sectors unit. When the reflector is in a horizontal position, this plane is vertical. In addition, the sectors are connected by two flat connecting pieces, arranged symmetrically on both sides of the plane of symmetry and at the intersections with the crossbars forming junctions (I, II, III, IV) with the intermediate structure. The journals are situated on the outer sides of the sectors on the extension of the axis of the pipe.

It is easy to see that the four supports formed by the rotating sectors unit have an identical rigidity in relation to the supporting journals as a result of the symmetry of the structure. This means that all four points (I, II, III, IV), when the reflector is in a horizontal position, being loaded by vertical unit forces, are imparted identical vertical movements; when the reflector is in a vertical position, they will be loaded by horizontal unit forces, directed antisymmetrically in relation to the horizontal axis of rotation of the reflector. This means that the mentioned points move in the direction of effect of the applied forces by the same amount

The intermediate eight-support structure unit is a latticework beam, consisting of four flat identical beams (to be called the main beams) 1-6, 2-5, 3-8, and 4-7 (Fig. 5), whose mutual intersection forms the square I, II, III, and IV, to the corresponding points of attachment of the intermediate structure to the rotating sectors. The cantilevers of the main beams are connected in pairs to the connecting beams 1-2, 3-4, etc., and the connecting shafts 2-3, 4-5, etc. The central square and the lateral rectangles, formed by the main beams, also contain the necessary connecting elements, shown by a dot-dash line. The points of intersection of the cantilevers of the main beams and the connecting beams form the eight principal fulcrums of the reflector: 1, 2, 3, . . . , 8.   These eight points lie in a single plane on a circle with a diameter of approximately $0.8D_{refl}$. When the reflector is in a horizontal position, this plane is horizontal.

It can be seen from the symmetry of the intermediate structure that when there is symmetrical loading (for example, from structural weight when the reflector is in a horizontal position), all eight supports have identical rigidity.

The eight-shaft pyramid (Fig. 6) with a central shaft forms the eight auxiliary fulcrums of the reflector, situated in the same plane on a circle whose diameter is approximately equal to $0.42D_{refl}$. When the reflector is in a horizontal position, this plane also is horizontal. The ninth supporting shaft passes through the axis of the pyramid. The vertex of the pyramid is attached to the midpoint of the counterweight beam.

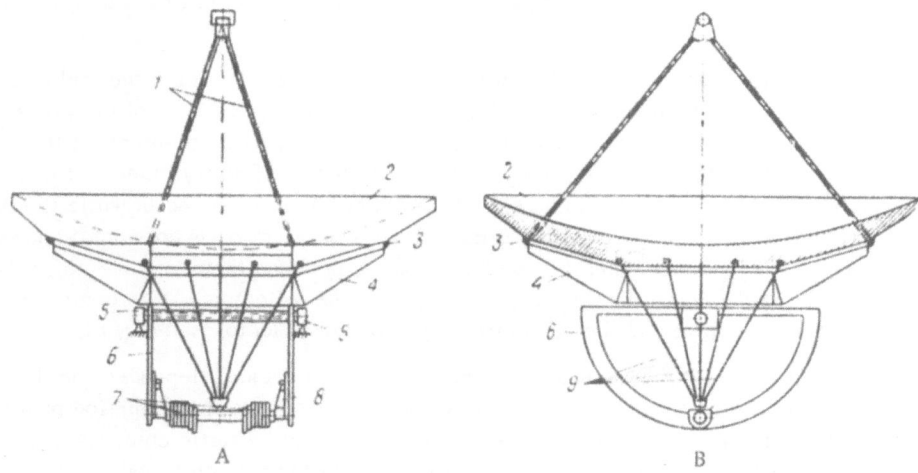

Fig. 6. Diagram of reflector suspension. 1) Four-shaft pyramid supporting the exciter system; 2) reflector; 3) reflector support; 4) intermediate supporting structure; 5) supporting bearing; 6) toothed sector; 7) counterweight; 8) lever for moving counterweight beam; 9) auxiliary eight-shaft supporting cone of reflector. A) Front view; B) side projection.

The framework of the reflector, with a radial design, consists of 16 radial flat beams, two dioctahedral circular (outer and supporting) beams, formed by the chord beams, a central support, several intermediate dioctahedral circular supports, and diagonal struts in the rear lower panels of the framework.

The eight principal and eight auxiliary fulcrums are arranged at the midpoints of the chord beams with two fulcrums in each even sector, formed by two adjacent radial beams. We will consider the first sector to be that whose axis is situated parallel to the horizontal axis of rotation. The central supporting shaft forms a fulcrum at the central point of the reflector.

With a horizontal position of the reflector, loaded by the structural weight, all eight principal fulcrums will have identical vertical movements relative to the supporting journals of the reflector due to the symmetry of design and load. The eight auxiliary fulcrums, formed by the eight-shaft pyramid, and also the central fulcrum, will experience identical vertical movements as a result of the identical rigidity of the pyramid shafts.

If identical rigidity of the intermediate eight-support structure and the eight-shaft pyramid with a central support (taking into account the rigidities of the pyramid shafts and the counterweight beam) is ensured, with a symmetrical loading all 17 fulcrums will experience identical movements, i.e., their relative movements along the direction of the loads will be equal to zero, which is equivalent to attachment of the reflector on 17 absolutely rigid supports or, in other words, the reflector is displaced without distortion of form, as an absolutely rigid body.*

In order to maintain the full symmetry of loads caused by structural weight, it is necessary to compensate the weight of the four-shaft pyramid supporting the exciter system. This pyramid is attached to the four principal fulcrums and the remaining four fulcrums should be loaded with corresponding ballast weight.†

Thus, when the reflector is in a horizontal position, the elastic deformations of the reflector caused by structural weight do not disrupt the initial (established during erection) arrangement of all 17 fulcrums and,

_____

*All 16 fulcrums move in radial directions, but these movements are insignificant due to the great rigidity of the reflector in radial directions.
† The radioastronomy laboratory of the FIAN now has developed a model of an eight-shaft pyramid.

therefore, the planes containing the eight principal and eight auxiliary fulcrums remain mutually parallel (and horizontal).

With rotation of the reflector about the horizontal axis, the plane containing the eight auxiliary fulcrums to all intents and purposes is distorted (as a result of the great longitudinal rigidity of the shafts of the eight-shaft pyramid in comparison with the flexural rigidity of the radial supporting members of the reflector frame-work), while the plane of the eight main fulcrums, under the influence of antisymmetrical loads, in a general case breaks down into two planes, one of which contains the fulcrums 1, 2, 5, and 6, while the other contains the fulcrums 3, 4, 7, and 8. In this case, the line of their mutual intersection is horizontal and is situated symmetrically in relation to the points 3-4 and 8-7, while the angle between them will be proportional to the component of the weight loads, parallel to the aperture plane of the reflector, i.e., proportional to the cosine of the angle between the horizontal and the geometrical axis of the reflector (see Fig. 5).

Which of the two planes will move ahead, and which will lag behind is dependent on the rigidities of the intermediate structure, the framework of the reflector, and the degree of participation of the reflector lattice and covering of the reflector in the work. This can be determined by static computations of the system made up of the rotating components, the intermediate structure, and the reflector, and can be checked using a model.

If the reflector was absolutely rigid, whatever the position in space, all eight fulcrums of the reflector and intermediate supporting structure would lie in a single plane. However, the reflector is not absolutely rigid and experiences elastic deformations under the influence of a load. The values of the elastic deformations are dependent on the modulus of elasticity E of the material and the values of the distribution and direction of the load and, therefore, on the form of the deformations (transverse flexure or buckling), on the support attachments, etc. Probably it would be possible to find such a relation of the rigidities of the supporting framework of the reflector and the intermediate supporting structure for which all eight fulcrums, being set in a single plane during erection which at that time was horizontal, would remain in this plane during the turning of the reflector in elevation angle.

However, the problem of finding such a relationship of rigidities in a multiply connected system is extremely unwieldy and it may have no solution because it cannot be asserted in advance that the sought-for relation exists. However, a different method for ensuring that all eight (principal) fulcrums will lie in a single plane during the turning of the reflector in elevation angle can be mentioned.

A three- or four-shaft pyramid, whose lower ends are attached to the reflector framework, and whose upper ends, forming the vertex of the pyramid, are situated beyond the focus of the reflector, usually is used for suspension of the system consisting of the exciter and its associated apparatus in the focus of the reflector.

The lower ends of the pyramid shafts can be attached to either the fulcrums 1-2 and 5-6 or 3-4 and 7-8. Whereas, in ordinary designs, the suspension of the exciter introduced additional deformations of the reflector, now the suspension instead is used for correction of deformations. With the turning of the reflector in elevation angle, the four-shaft pyramid supporting the exciter system imparts a moment to the intermediate supporting structure. Naturally, the lower ends of the pyramid shafts should be attached to those four fulcrums whose plane lags during the turning of the reflector. The value of the moment can be adjusted by the weight which is situated at the vertex of the pyramid or beyond it and its shoulder. When the reflector is in a horizontal position, the moment from the pyramid is equal to zero. The variation of the moment during turning of the reflector occurs in conformity to the same law as the variation of the skew-symmetric components of loads caused by structural weight.*

----

*In the case of quasi-four- and quasi-eight-shaft pyramids supporting the exciter system (see preceding footnote), the presence of the eight principal fulcrums in a plane is ensured by the longitudinal rigidity of the shafts of this pyramid. The use of a special design of the shafts ensures insignificant additional shading of the reflector.

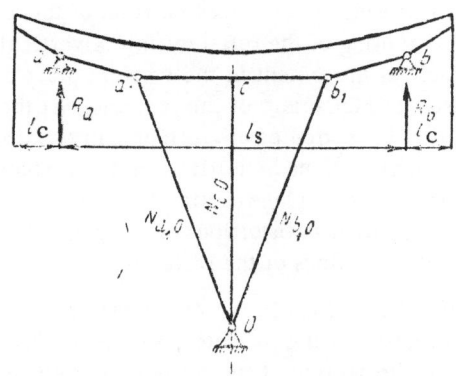

Fig. 7. Diagram of diametral section of reflector.

Suspension of a reflector using an intermediate eight-support structure has one other significant advantage. Experimental checking of deformations of the reflector after completion of erection or in the course of operation, regardless of position, is a complex problem. A determination as to whether the reflector has retained its parabolic form is made indirectly, on the basis of its electrical parameters.

When there is an intermediate supporting structure, an appraisal of the form of the reflector can be made using the following method.

During the adjustment of the intermediate supporting structure to which the reflector is attached in a horizontal position, a leveling instrument is used for determining the readings for eight support points of intersection lying in a single plane. By turning the reflector in elevation angle until it is in a vertical position, it is possible to check (using a theodolite) whether all eight fulcrums have remained in the same plane. If there has been disruption of the position of the fulcrums relative to the plane, it is necessary to take the measures mentioned above (weight balancing); if there has been no disruption, the distortion of the form of the reflector will be within the limits of the known computed value in the spans and cantilevers of the reflector (this is confirmed by model investigation of deformations).

Such a "plane" checking also can be made with the reflector in an intermediate inclined position. In this case the theodolite must be set in the corresponding inclined plane.

Functioning of an Eight-Shaft Pyramid with the Reflector in Horizontal and Vertical Positions

Figure 7 shows a diametral section of the reflector, where a and b are the principal supports, a_1 and b_1 are the auxiliary supports, l_c are the cantilever parts, and l_s are the spans. The auxiliary supports a_1, b_1, and c have the long shafts a_1O, b_1O, and cO.

The lengths of the pyramid shafts are adjustable, making it possible to set all the auxiliary fulcrums in a single plane. Assuming all the supports to be absolutely rigid (for the auxiliary supports this is the vertex O of the shaft pyramid), in the first approximation we determine the excess unknown stresses in three shafts: N_{a_1O}, N_{b_1O}, and N_{cO}, and the reactions of the principal supports R_a and R_b. For a more precise solution, it also is necessary to take into account the circular elements of the reflector framework, i.e., the chord beams.

The eight auxiliary fulcrums, like the principal fulcrums, are arranged in a radial-symmetric pattern, i.e., they lie on a single plane surface whose diameter is approximately half the diameter of the circle of main fulcrums. The ninth support point is for the central support which passes along the geometrical axis of the reflector. The vertex of the pyramid (with a central support) is attached to the counterbalance beam, connecting the lower points of the rotating sectors.

If we know the loads on the principal supports we can determine their vertical displacement relative to the supporting journals of the reflector. By knowing the stresses in the shafts of a nine-shaft structure, we can determine the vertical movement of the point O, equal to the flexure of the counterbalance beam plus the movements of the supporting ends of the counterweight beam (these occur due to deformations of the rotating sectors).

The vertical movements of the principal fulcrums should be equal to the vertical movements of the auxiliary fulcrums (with buckling of the pyramid shafts taken into account); this is ensured by an appropriate selection of the flexural rigidity of the counterbalance beam and the cross section of the pyramid shafts.

When these conditions are observed, the reflector will have 17 fulcrums in a radial-symmetric pattern. Figure 8 shows a diagram of the arrangement of the 17 fulcrums as viewed from above; Fig. 9 shows a diagram of the system as a whole.

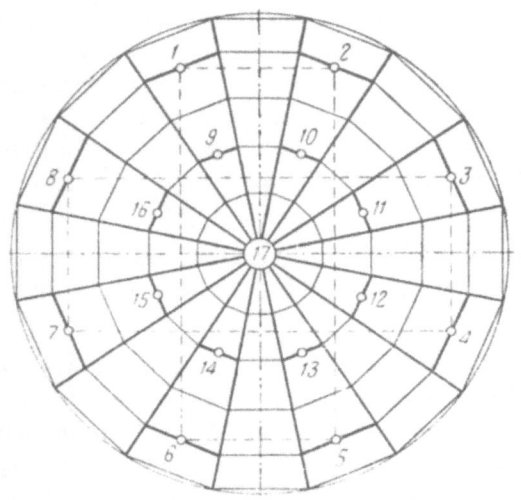

Fig. 8. Diagram of arrangement of 17 reflector fulcrums. 1,2,3,...,8) Principal fulcrums; 9,10, ...,17) auxiliary fulcrums.

Now we will consider the vertical position of the reflector. First, upon turning of the reflector into a vertical position, it experiences movement in the vertical plane as a result of elastic displacement of the supports. At the same time, the circular section of the reflector in relation to the principal supports retains its initial true form (since the form of the circle and its plane is maintained). Second, there will be skew-symmetric distortions of the parabolic curve in the diametral sections of the reflector.

The solid line KK in Fig. 10 represents the undeformed diametral section of the reflector, and the dashed line K'K' represents the section of the reflector deformed under the influence of skew-symmetric loads caused by structural weight.

We will assume that the vertex of the eight-shaft pyramid is not attached. Then, being sort of an appendage of the framework of the reflector and being rigidly connected to it, it will move from the point O to the point O' and assume the position shown in Fig. 10 by dashed lines (the ninth, central, shaft is not affected by skew-symmetric loads).

In order to simplify reasoning, for the time being we will consider the eight-shaft pyramid to be weightless. Due to the symmetry of design and the antisymmetry of the load, the absolute values of movement of

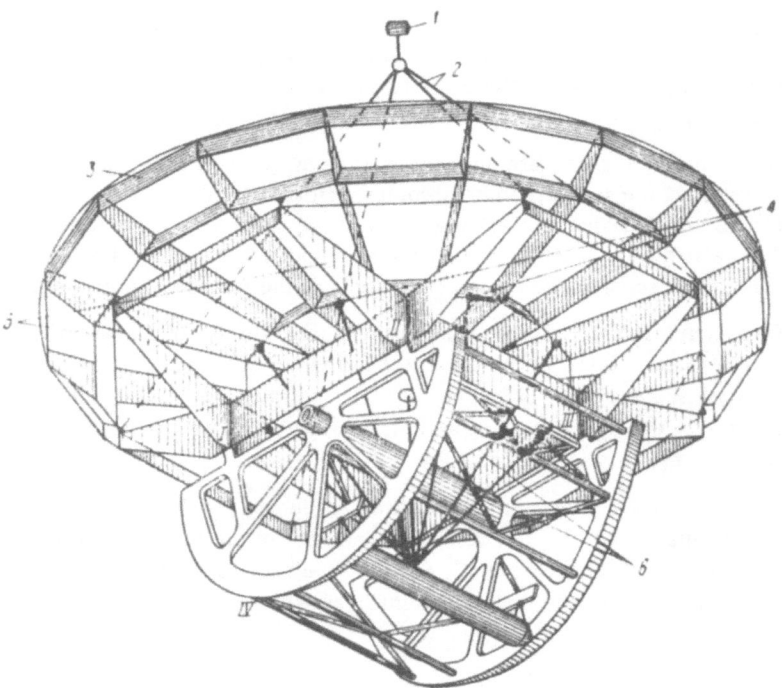

Fig. 9. Model of reflector with intermediate supporting structure, rotating sectors, and ten-shaft pyramid. 1) Exciter counterweight; 2) four-shaft pyramid; 3) reflector framework; 4) intermediate fulcrums (9, 10,...,16 in Fig. 8); 5) intermediate eight-support structure; 6) eight-shaft pyramid.

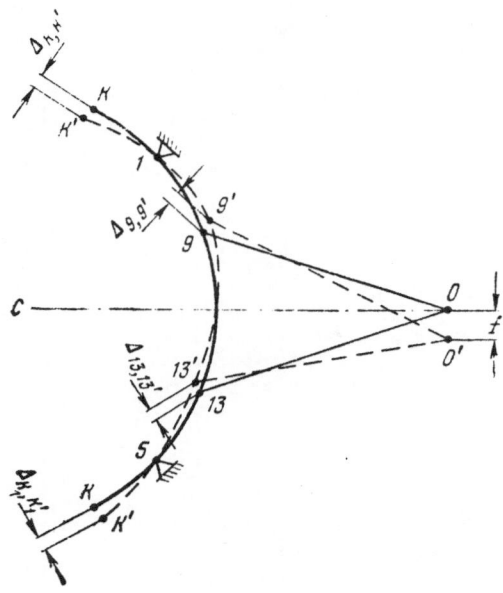

Fig. 10. Diagram of deformations of reflector.

symmetrical points will be equal to one another:

$$| \Delta_{K,\ K'}| = |\Delta_{K_1,\ K_1'}|;\ |\Delta_{9,9'}| = |\Delta_{13,13'}|.$$

However, in actuality, the vertex of the eight-shaft pyramid is not free but is attached to the counterweight beam. The attachment of the pyramid vertex prevents its movement to the point O' and, therefore, prevents deformation of the parabolic curve of the reflector.

Theoretically, it would be possible to compute the vertical movement of the principal supports of the reflector and the values of the buckling $\Delta_{9,9'}$ and $\Delta_{13,13'}$ and accordingly select the flexural rigidity of the counterweight beam, also varying the balance weights, in order to ensure the required vertical movement f of the vertex of the pyramid. However, it is better to do this by a different method, more reliable and clear-cut. On the basis of the qualitative picture of deformations (Fig. 10), it can be concluded that for compensation of the deformations distorting the parabolic form of the reflector, it is necessary to impart by force vertical transverse movements to the vertex O of the pyramid.

Balance-Kinematic Compensation of Skew-Symmetric Deformations of Reflector from Loads of Structural Weight

The shape of the deformed curve of the diametral section of the reflector (dashed curve), shown in Fig. 10, resulting from the effect of skew-symmetric loads, applies to any diametral section of the reflector. The absolute values of the movements $\Delta_{K,K'}$ and $\Delta_{9,9'}$ are proportional to the cosine of the angle of inclination of the diametral section to the vertical. The movements of the end points of the shafts of the octahedral pyramid, forming the eight intermediate supports, with the turning of the axis OC of the pyramid (Fig. 10) in the vertical plane, also are proportional to the cosine of the angle of inclination of the section in which the particular intermediate support is situated.

Therefore, by imparting a corresponding forced movement to the vertex of the pyramid in the vertical plane (with elastic deformation of the pyramid shafts taken into account), it is possible almost completely to compensate the skew-symmetric deformations of the reflector framework (with an accuracy to the free play of the ball-and-socket connections at intersections).

The FIAN radio astronomy laboratory has developed a design for a mechanism used in balance-kinematic compensation of skew-symmetric deformations of the reflector caused by structural weight. The kinematic compensation mechanism consists of the following elements (Fig. 11): a large-toothed rim (1) attached to the housing of the reflector supporting bearing (2); a small gear (3) which meshes with the toothed rim; two conical (4) and one cylindrical (5) toothed pair; a transmission shaft (6), a coupling box, permitting angular deflection and reciprocal linear displacements; an eccentric (7) with adjustable eccentricity and a lever-dog system (8); a counterbalance beam (9); joint (10); spherical pivot (11); and the counterweight collars (12).

The upper conical toothed pair, transmission shaft, lower conical toothed pair, and cylindrical toothed pair are located within the metal structure of the rotating sector (the rotating sector is not shown in the figure). The letter O denotes the point of attachment of the vertex of the pyramid and the arrows C_1 and C_2 indicate its movements. The mechanism consists of two symmetrical segments for greater rigidity and accuracy.

The figure shows only the left segment of the mechanism; the right segment has not been shown for simplification of the figure. As the reflector is turned in elevation angle, the rotating sector draws along the axis of the small gear and the conical wheel. The small gear, meshing with the toothed rim, is imparted rotational motion.

Fig. 11. Diagram of balance-kinematic compensation mechanism.

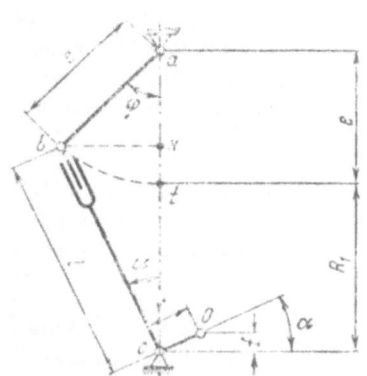

Fig. 12. Diagram of connection of eccentric to lever of counter-weight beam.

By means of the above-mentioned transmission links, rotation is imparted to the eccentric, which is situated on the shaft of the last cylindrical toothed wheel. The gear ratio from the fixed toothed wheel to the shaft of the last cylindrical wheel on which the eccentric is situated is equal to unity.

Thus, the shaft of the eccentric duplicates the turning of the reflector and, in turn, by means of the lever-dog, brings the counterweight beam into motion about its axis. The turning of the eccentric, as well as the reflector, is restricted to ±90°. Since the arm R_1 of the lever is many times greater than the eccentricity of the eccentric (in the design it was assumed that R_1/l = 40), the turning of the counterweight beam falls in the range ±1.5°. The lateral displacements of the points of attachment of the pyramid vertex (Fig. 11, arrows C_1 and C_2) to all intents and purposes occur along a straight line (along an arc whose central angle is ±1°.5).

Now we will consider the change of the value of lateral displacement of the point of attachment of the pyramid vertex. Figure 12 is a diagram of the connection of the eccentric to the lever of the counterbalance beam. The link ab is connected rigidly to the reflector and the angle φ is the angle of rotation of the reflector relative to the horizontal axis.

Using the notations in Fig. 12, we have

$$bk = e \sin \varphi = l \sin a,$$

hence,

$$\sin \alpha = \frac{e}{l} \sin \varphi. \tag{18}$$

From the triangle abc,

$$l = \sqrt{e^2 + (e + R_1)^2 - 2e(e + R_1)\cos\varphi} = e\sqrt{2 + 2\frac{R_1}{e} + \left(\frac{R_1}{e}\right)^2 - 2\cos\varphi - \frac{R_1}{e}\cos\varphi}.$$

Substituting the value l into (18), we obtain

$$\sin\alpha = \frac{\sin\varphi}{\sqrt{2 + 2\left(\frac{R_1}{e}\right) + \left(\frac{R_1}{e}\right)^2 - 2\cos\varphi - 2\frac{R_1}{e}\cos\varphi}};$$

$$f = r\sin\alpha = \frac{r\sin\varphi}{\sqrt{2 + 2\left(\frac{R_1}{e}\right) + \left(\frac{R_1}{e}\right)^2 - 2\cos\varphi - 2\frac{R_1}{e}\cos\varphi}}.$$

(19)

For the assumed relation $R_1/e = 40$, we have

$$2\left(\frac{R_1}{e}\right) = 80; \quad \left(\frac{R_1}{e}\right)^2 = 1600;$$

$$2\left(1 + \frac{R_1}{e}\right)\cos\varphi = 82\cos\varphi; \quad \sin\alpha = \frac{\sin\varphi}{\sqrt{1682 - 82\cos\varphi}};$$

$$f = r\frac{\sin\varphi}{\sqrt{1682 - 82\cos\varphi}}.$$

(20)

It was mentioned above that the skew-symmetric deformations of the reflector are proportional to the skew-symmetric component of the loads from the weight of the apparatus (i.e., the component parallel to the aperture plane of the reflector). The right-hand side of (20) differs somewhat from the sine function, but this deviation is insignificant since the value $\sqrt{1682 - 82\cos\varphi}$ with a change of φ from 0 to 90° varies in the range

$$\sqrt{1600} < \sqrt{1682 - 82\cos\varphi} < \sqrt{1682}.$$

The maximum error of deviation from the sine function is less than 5%.

Should it prove necessary, the error can be reduced still further by increasing the ratio R_1/l.

It can be seen from relation (19) that the value of the lateral displacement is dependent on eccentricity e: when e = 0, $\sin\alpha = 0$ and $f = 0$.

Thus, in the adjustment process, the compensation of the skew-symmetric flexures of the reflector can be regulated in accordance with the actual deformations. The criteria of successful compensation in this case are: a) maintaining all eight auxiliary fulcrums in a plane, and, b) maintaining the planes of the auxiliary and principal fulcrums parallel.

It was mentioned earlier that in principle the lateral movements of the vertex of an eight-shaft pyramid could be brought about simply by elastic deformations of the counterweight beam, also varying the balance weights.

This possibility also is used in the balance-kinematic compensation method. The fact is that under the influence of the stresses in the links of the kinematic compensation mechanism there will be elastic deformations and in the kinematic pairs there will be elastic free play.

This phenomenon can be decreased considerably if balance weights are used for creating the required torque on the beam.

The balancing weight (the counterweight weight balancing the reflector relative to the horizontal axis of rotation) is not all within the counterweight beam, but instead in part sits on the beam in the form of reinforced concrete collars (see Fig. 6). The collars are removable and the openings are in an eccentric position, so that

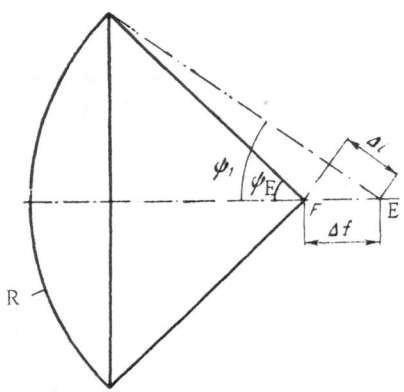

Fig. 13. Diagram of displacement of the exciter in the reflector focus.

the center of gravity of the collar is displaced relative to the axis of the counterbalance beam. If, in the case of an even number of collars, half of them are attached to the beam to the right of the center of gravity (looking along the axis of the beam when the reflector is in a horizontal position) and half are attached to the left, the resulting torques, added together, will be equal to zero regardless of the position of the reflector. However, if (when the reflector is horizontal) the centers of gravity of all (or part) of the collars are arranged in such a way that they are below the axis of the counterbalance beam and fall in the vertical plane, with turning of the reflector in elevation angle a torque arises on the counterbalance beam which is directed in such a way that the vertex of the eight-shaft pyramid is moved upward. By varying the value of the torque created by the weighting collars, it is possible to facilitate the operation of the kinematic compensation mechanism, which now must overcome only the difference between the necessary torque and the torque created by the balancing collars. With such joint operation of the mechanism and the balancing weights, the stresses in the links of the mechanism, and therefore the deformations as well, can be reduced to a minimum.

Design of Exciter Supporting System

A three- or four-shaft pyramid usually is used for suspending the exciter and its high-frequency components in the reflector focus. The satisfaction of the requirements imposed on the rigidity of suspension of the exciter system is not the least important of the problems to be solved when creating parabolic antennas of large size.

A reflector designed for operating on short waves should have a relatively large focal length for optimum radiation*

$$F \approx (0.4 - 0.60) \, D_{\text{refl}},$$

where F is the focal length and D_{refl} is the diameter of the reflector. For example, for D_{refl} = 66 m, it is necessary that F be about 40 m. The shafts of the four-shaft pyramid supporting the exciter system will be of approximately the same length.

If the shafts are of considerable length, the shafts of the pyramid should have the smallest possible transverse dimensions in order to lessen the shading of the reflector; they also should be as light as possible in order to decrease the load on the reflector. Therefore, the possibilities of increasing the rigidity of the shafts are limited. The deformations of the shafts of the pyramid from the weight of the structure and the weight of the apparatus of the exciter system are limited by the requirement of minimum aperture field phase distortion [4].

If f is the displacement of the phase center of the exciter E (Fig. 13) from the focus F of the reflector R, the increase of the length of the beam path at the center of the reflector will be equal to f, and at the edge of the reflector

$$\Delta l = \Delta f \cos \psi_1 \approx \Delta f \cos \psi_0$$

(in the case of small displacements, Δf).

The maximum difference in the increase of the length of the beam path will be between the beam reflected from the center of the reflector and the beam reflected from its edge,

$$\Delta f - \Delta l = \Delta f - \Delta f \cos \psi_0 = \Delta f \, (1 - \cos \psi_0).$$

---

* The conditions of decrease of noise temperature require a somewhat smaller F/D_{refl} ratio:

$$F/D_{\text{refl}} \approx (0.4 - 0.5).$$

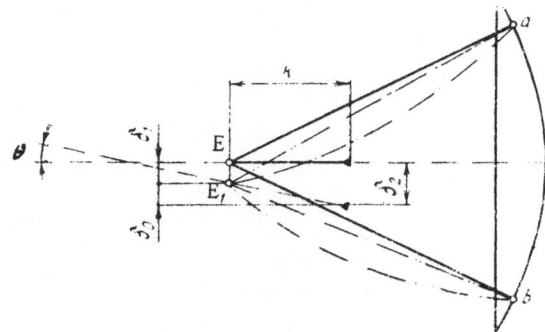

Fig. 14. Diagram of deformations of the system supporting the exciter.

The phase change corresponding to this will be

$$\Delta\alpha = \frac{2\pi}{\lambda}\,\Delta f\,(1 - \cos\psi_0).$$

Assuming the value of the admissible phase error to be $\Delta\alpha = \pi/4$, we obtain

$$\Delta f_{\max} = \frac{\lambda}{8\,(1 - \cos\psi_0)}. \qquad (21)$$

Relation (21) shows that the admissible displacement of the exciter from the reflector focus increases for a given diameter with an increase of focal length (with a decrease of the aperture angle $2\psi_0$).

For the 22-m RT-22 reflector of the FIAN, having $2\psi_0 = 120°$, the admissible error of placement of the exciter is

$$\Delta f = \frac{\lambda}{8\,(1 - \cos 60°)} = \frac{\lambda}{8 \cdot 0.5} = 0.25\,\lambda.$$

For example, when $\lambda = 8$ mm, $\Delta f = 2$ mm. For the 25-m Bonn radio telescope, having $2\psi_0 \approx 140°$,

$$\Delta f = \frac{\lambda}{8\,(1 - \cos 70°)} = \frac{\lambda}{8\,(1 - 0.34)} \approx \frac{\lambda}{5.3}.$$

The admissible value of the lateral displacement of the exciter from the focus of the reflector is governed by the requirements of maintaining the direction of the electrical axis and also the coefficient of directional effect [4]. The deviation of the direction of the electrical axis from the direction of the optical (geometrical) axis of the reflector, depending on the lateral displacement of the exciter, is less in the case of reflectors with long focal length than for reflectors with short focal length. For the 22-m RT-22 reflector with a focal length $F = 9.525$ m and $R = 11$ m, the angle of deviation of the electrical axis is

$$\theta_2 \approx 0.8\theta_1,$$

where θ_1 is the angular value of the lateral displacement of the exciter.

With displacement of the exciter by 2 mm,

$$\theta_2 \approx \mathrm{tg}\,\theta_2 = \frac{2}{9525} \approx 40''.$$

Since the determined width of the directional diagram at half-power for this reflector at $\lambda = 8$ mm is about 2' [1], a deviation of the electrical axis $\theta_2 \approx 40''$ can be considered acceptable, although it is of the same order of magnitude as the width of the diagram for this wavelength.

The displacements mentioned vary regularly with elevation angle and, therefore, it is possible to introduce corresponding corrections for the radio telescope adjustment. The decrease of the coefficient of directional effect does not exceed 1-2%, which can be considered acceptable.

The exciter should be as distant as possible from the point of intersection of the four shafts (i.e., from the vertex of the pyramid) in order to decrease the influence of metal parts on the operation of the exciter and also for decreasing the shading of the reflector by the exciter supporting system. This circumstance leads to still greater lateral displacements of the exciter from the focus of the reflector when the latter is turned in elevation angle.

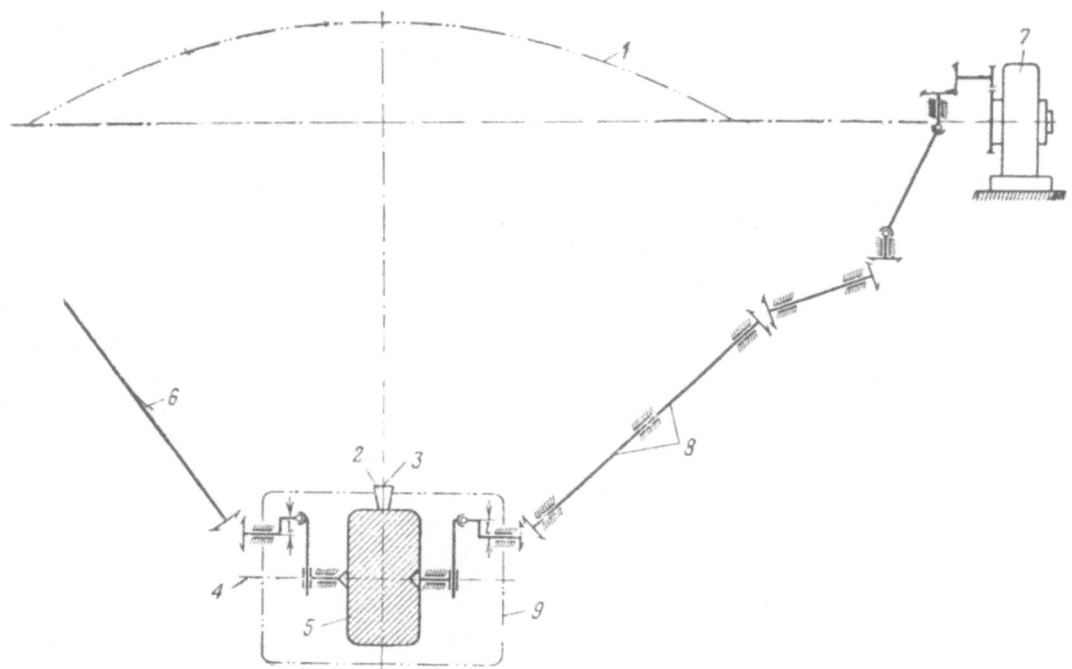

Fig. 15. Diagram of the kinematic compensation mechanism. 1) Reflecting surface of parabolic reflector; 2) exciter horn; 3) reflector focus; 4) axis of rotation of reflector; 5) framework with exciter, suspended on horizontal axis; 6) second symmetrical segment of kinematic compensation mechanism; 7) supporting bearing of reflector; 8) transmission shaft, situated within the supporting shaft of a four-shaft pyramid; 9) cabin with radio receiving apparatus (high-frequency components and exciter), situated beyond the focus of the reflector. The eccentricity l is regulated during adjustment.

When the reflector is in a horizontal position, the exciter moves only along the geometrical axis of the reflector. This movement, usually insignificant, is governed by the buckling of the pyramid shafts under the load of the reflector and exciter. However, when the reflector is in a vertical position, the displacement of the exciter (which now is lateral relative to the geometrical axis of the reflector) occurs due to buckling and flexural deformation of the shafts, with the influence of the latter component being many times greater than the first.

Figure 14 is a diagram of the supporting structure (four-shaft pyramid) for the exciter system when the reflector is in a vertical position. The dashed lines show the deformed state of the shafts.

Using the notations in Fig. 14, we have

$$\delta_2 = \delta_1 + \delta_0 = \delta_1 + K \operatorname{tg} \theta \approx \delta_1 + K\theta. \tag{22}$$

Computations show that the movements of the exciter of a 66-m parabolic reflector in a horizontal position are equal to $\delta = 1$ mm, and in a vertical position, $\delta = 1.5$ mm, $\delta_0 = 29.5$ mm, and $\delta_2 = 31$ mm.

The displacements of the exciter from flexure of the shafts could be compensated using a counterweight placed beyond the vertex of the four-shaft pyramid (to the left of point O in Fig. 14), but it is impossible to completely compensate the displacement of the exciter by this method because the additional load causes an increase of the first displacement component (δ_1). In addition, use of compensation of displacement of the exciter by a load moment involves an increase of the loads on the reflector and additional deformations.

For a reflector with a diameter of 66 m, the counterweight would be approximately 10 tons.

The radio astronomy laboratory of the FIAN has devised a method for kinematic compensation of displacement of the exciter occurring due to weight of the structure. This method in essence is a modification of the method for balance-kinematic compensation of skew-symmetrical deformations of a reflector, but without use of balance loads, since the stresses in this case are insignificant.

The exciter and the high-frequency components are mounted in a special framework which is suspended in two supporting bearings, forming a horizontal axis. The framework can rotate within the necessary range about this axis. The framework with the exciter is turned about the horizontal axis by means of the kinematic compensation system. Figure 15 is a diagram of the mechanism for kinematic compensation of displacements of the exciter. The mechanism is made from two symmetrical segments, operating in parallel, for high accuracy and reliability. The left segment is not shown in the figure.

Despite the great length of the transmitting links of the mechanism, the total length of the transmission shafts for a reflector with D_{refl} = 66 m is about 60 m in one segment. The torque on the entire length of the shaft (elastic deformation) is approximately 30'.

The use of kinematic compensation of displacement of the exciter solves the problem of the admissible deformation of the shafts of the supporting structure of the exciter resulting from weight loads. Of course, the flexural rigidity of very long shafts must satisfy other requirements, such as resistance to buckling and absence of resonance phenomena accompanying wind loads.

Temperature Deformations of a Parabolic Reflector

Until now we have assumed that the principal and constantly operating loads causing elastic deformations of a reflector are the weight of reflector and the wind; when the wind velocity is about 12 m/sec, the wind loads are about 10% of the loads from the weight of the reflector. Wind loads have been inadequately studied, but nevertheless their values and distribution for the components of a reflector are known sufficiently for computations in the first approximation. It can be assumed, as mentioned above, that if the elastic deformations of the reflector caused by its own weight do not exceed the stipulated values, the elastic deformations from wind loads, which are only 10-12% of those caused by the weight of the reflector, in case of necessity can be compensated by a corresponding increase in the rigidity of the reflector without risk of an excessive increase of the weight of the reflector (especially in the case of a radial-symmetric multisupport design of the suspension).

As mentioned at the beginning of the article, under the actual conditions for use of parabolic antennas there also will be temperature deformations, caused by change of the temperature of the components of the reflector and the coefficient of linear expansion of the material used in its construction.

The influence of temperature on the form of a parabolic reflector in the case of uniform heating is considered in [8, 10]. However, in the case of uniform heating, the temperature deformations are less dangerous because, in such cases, in the first approximation there is only a change of the focal length of the paraboloid.

We will explain this by simple computations. The equation for the parabola forming the paraboloid of revolution for a given temperature (t_0) has the form (Fig. 16)

$$y_0 = \frac{x_0^2}{4F}. \tag{23}$$

With a change of temperature by Δt, the point a_0 is moved to a_1:

$$y_1 = y_0 (1 + \alpha \Delta t);$$

$$x_1 = x_0 (1 + \alpha \Delta t). \tag{24}$$

Substituting the values x_0 and y_0 from (24) into Eq. (23), we have

$$y_1 = \frac{x_1^2}{4F (1 + \alpha \Delta t)}, \tag{25}$$

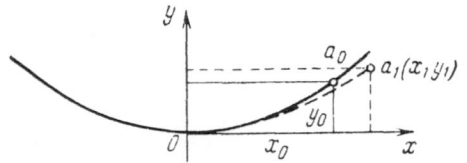

Fig. 16. Deformation of a paraboloid in the case of uniform change of heating temperature.

that is, we obtain a parabola with the focal length

$$F_1 = F (1 + \alpha \Delta t). \tag{26}$$

If we denote the lengths of the shafts of a four-shaft pyramid supporting the exciter at a given temperature by l_0, with a change of temperature by Δt, the lengths of the shafts will be

$$l_1 = l_0 (1 + \alpha \Delta t), \tag{27}$$

that is, they change in the same ratio as the focal length of the reflector and the exciter will not go out of focus. With respect to nonuniform heating of the components of the reflector, as far as is now known to the author, there are insufficient experimental data available on this problem. Possibly, by making temperature measurements at many points of the reflector during its operation, on the basis of recorded data it would be possible to compute the temperature deformations of the reflector for a particular temperature distribution. However, it is easiest to solve this problem experimentally. For this it is necessary to orient the reflector in a horizontal position and use a leveling instrument to check the plane distribution of the control points of the reflector — the principal and auxiliary fulcrums.

After obtaining data on the values of displacement of the fulcrums from their plane, and knowing the distribution of temperatures for the components of the reflector and its suspension in this way, by analyzing the results of temperature and flexure measurements of the control points, conclusions can be drawn on the dependence of the values of the actual deformations of the reflector on nonuniform heating.

Depending on the results of such an investigation, which makes it possible to estimate the absolute values of the temperature deformations, it will be possible to make sound decisions concerning the methods to be used for their compensation.

In addition to painting the components of the reflector and suspension white, it is desirable that heat insulation be used. The fact is that in seeking to eliminate temperature deformations there is no need at all to maintain any definite temperature of the entire telescope; it is only necessary to eliminate or at least decrease the temperature difference of its individual components.

A four-shaft pyramid, supporting the exciter system and incorporating heat insulation, has been developed at the radio astronomy laboratory of the FIAN. Each of the four shafts is made in the form of a thin-walled tube (a cylindrical casing with diaphragms imparting additional rigidity, covered by a layer of white foam plastic 10 mm thick).

The use of heat insulation for the shafts is desirable even in a case when the entire remaining part of the reflector is not safeguarded against nonuniform heating. The surface of the reflector is smoothly articulated and there are no significant temperature gradients along it, whereas the shafts of the pyramid, arranged at different angles to the solar rays, may differ appreciably in temperature. On the other hand, even an insignificant difference in the temperatures of heating of the shafts leads to great displacements of the exciter because of their great length.

For example, in the case of a temperature difference between two opposite pairs of shafts of only 5°, the difference in the lengths of the shafts will be equal (for a reflector with D_{refl} = 66 m) to

$$\Delta l = l \alpha \Delta t = 40,000 \cdot 12 \cdot 10^{-6} \cdot 5 = 2.4 \text{ mm},$$

and the lateral displacement of the exciter, with an angle between the axes of shafts with different temperatures $\alpha / 2 \approx 40°$, is

$$\delta_2 = \frac{\Delta l}{\cos \alpha/2} = \frac{2.4}{0.766} \approx 3 \text{ mm}.$$

There are at least four additional ways to decrease temperature deformations:

1. Use of servomechanisms for compensation of thermal deformations by mechanical movements of the components of the reflector.
2. Use of electrical heating and cooling of the components.
3. Construction of reflector components from a ferronickel alloy with a small coefficient of linear expansion, such as invar or kovar.
4. Construction of protective domes from materials transparent to radio waves.

Without discussing the details of each of these solutions, we note only that adoption of these various measures, which would involve great additional expenditures, can be justified only in a case where study of the problem of temperature deformations reveals the desirability of such measures.

The designs described, making it possible to solve the problem of small deformation for a parabolic antenna of considerable size (resulting from weight loads), served as the basis for drawing up plans for a radio telescope with a fully steerable 60-m reflector. As an illustration of the basic principles formulated, we will give a brief description of some of the components developed in the working plans for the 60-m reflector.

Design of Individual Units of the Reflector and Its Suspension

The design of individual units of the 60-m parabolic reflector was developed in the radio astronomy laboratory of the FIAN. The reflector was suspended at 17 points with a radial-symmetric arrangement relative to its framework.

The individual components of the structure were developed for the purpose of finding possible successful designs and determining specific cross sections for the supporting components for use in static computations. These components will be described.

Unit of Rotating Sectors with Horizontal Axis and Counterweight Beam. The unit is useful principally because: a) it should ensure conversion from a two-support suspension of the reflector to a four-support suspension; b) it performs the role of the last transmission stage (toothed sectors) of the mechanism for turning the reflector about its horizontal axis. Taking this into account, the unit of rotating sectors with respect to supporting properties should possess great flexural rigidity in relation to the two supporting journals and torsional rigidity in the planes parallel to the planes of the rotating sectors.

This unit consists of two rotating sectors (diameter of the initial circle $D_{i.c.}$ = 15,800 mm), made in the form of a multiply connected framework of welded sheet steel of the thickness δ_{sh} = 10-40 mm; a rolled steel pipe 3000 mm in diameter, whose axis is the horizontal axis of rotation of the reflector; a counterbalance beam, also made in the form of a rolled pipe 2300 mm in diameter, filled with concrete; eccentric weight collars, made in the form of closed casings, filled with concrete; lateral tubular connecting pieces, arranged around the circle of the rotating sectors near the rim normal to the planes of the sectors, in the form of a squirrel cage; diagonal tubular struts, together with the lateral connecting pieces and the rotating sectors forming a spatially rigid structure capable of absorbing large flexural moments and torques; and two tubular braces, suspending the counterweight beam in the plane passing through the horizontal axis of rotation and the axis of the counterweight beam.

The toothed rims of the rotating sectors are made as toothed rods, curved smoothly along the initial circle on rollers (with linings) and set on both sectors using studs made in the field as required by the particular radius.*

---

* The studs are fabricated during the erection work when the sectors are mounted in the planned position on the supporting bearings. The rotating sectors unit as a whole is rotated using winches and cables and is secured rigidly in position by temporary tie-rods when each stud approaches the head of the grinding lathe set up alongside each sector.

The horizontal pipe, whose axis is the horizontal axis of rotation of the reflector, is connected rigidly to the sectors by bolts on the flanges. The flanges of the pipe and the hubs of the sectors are formed mechanically. On the outer sides of the sectors there are journals which are bolted to the sectors and to the flanges of the pipe. The journals have spherical rims and together with twelve-roller bearings ensure self-adjusting spherical supports.

The counterbalance beam has spherical endings to serve as supports. The housing of the supporting bearings, containing split spherical bushings matching the spherical endings of the pipe, are mounted within the sectors with a center at the middle of the plane in which the sectors lie. At the ends of the counterbalance beam there are levers used in turning the beam about its axis.

The beam carries eccentric weighting collars which are arranged symmetrically to the midpoint of the beam in such a way that when the reflector is turned in elevation angle about its horizontal axis, the weight collars create a torque on the beam equal to

$$M_{tor} = eG_{col} \sin \varphi, \tag{28}$$

where e are the eccentric collars, G_{col} is the weight of the collars, φ is the angle of rotation of the reflector relative to the horizontal axis (the axis between the geometrical axis of the reflector and the vertical). In the midpart of the counterbalance beam there is a rib with an opening for ball-and-joint attachment of the vertex of the octahedral shaft pyramid, forming eight auxiliary supports of the reflector.

Balance-Kinematic Compensation Mechanism. As already mentioned above, the balance-kinematic compensation mechanism serves for a control of the eight auxiliary fulcrums of the reflector, i.e., for ensuring that the plane in which the eight auxiliary fulcrums lie, and the plane of the eight principal fulcrums, remain parallel to one another by means of lateral movement of the vertex of the shaft pyramid.

The balance-kinematic compensation mechanism consists of two parallel segments, each of which has a large fixed toothed wheel which is attached to the corresponding housing of the supporting bearing, and a small satellite which is attached to the sector framework and which together with it moves about the fixed toothed rim. On the same axis with the satellite there is a conical toothed wheel which meshes with the toothed wheel seated on the end of the transmission shaft. At the other end of the transmission shaft there is a second conical toothed wheel which meshes with the wheel seated on the same axis with the cylindrical toothed wheel which imparts motion to the eccentric of the counterbalance beam and lever. The entire mechanism is situated within the framework of the sector. The free-play in the bearings and toothed meshing is determined by trial and error; the efficiency of the mechanism, which is greatly lowered because of this factor, is not important enough to consider, because the drive for rotating the reflector has reserve power for overcoming the moment from wind loads.

Intermediate Eight-Support Structure. As already mentioned, the intermediate eight-support structure is used for converting from a four-support suspension of the reflector to an eight-support suspension. Accordingly, it is designed in such a way that it has four units for connection with the corresponding four units in the rotating sectors and forms the eight principal supports of the reflector. The intermediate eight-support structure includes four main identical flat connecting pieces made from tubular strips and I-struts which are connected securely to the strips; four flat end connecting pieces, also made from pipes and I-beams; internal beam crosspieces, made from I-beams; peripheral tubular bracings and diagonal tubular struts in panels, formed by the main end connecting pieces.

All flat connecting pieces are welded but they are bolted into the three-dimensional structure, so that they can be removed. The end braces, of the main flat connecting pieces are made from $\phi\, 560 \times 12$ pipes. The upper flanged ends of these braces, surrounded by circular ribs and filled with babbitt, form surfaces to which are attached the bases of the split housings of the spherical bearings, being the principal supports of the reflector. The surfaces formed by babbitt are smoothed and are determined so as to lie in the same plane (all eight surfaces) by using a leveling instrument.

Octahedral Shaft Pyramid with Central Support. The octahedral shaft pyramid with a central support is made up of eight tubular shafts connected at the lower point (vertex of the pyramid), with bolting of the unit along the flanges. The upper ends of the tubular shafts have conical endings with forks, forming the eight auxiliary supports of the reflector. The length of the shafts is regulated in such a way that all eight auxiliary fulcrums will be in the same plane, parallel to the plane of the eight principal fulcrums of the reflector.

Reflector. The framework of the reflector is radial-symmetric. It consists of 16 radial connecting pieces, a central support, and 10 circular dioctahedral connecting pieces, each consisting of 16 flat chord connecting pieces. The lower panels of the framework, formed by the lower linear strips of the radial and chord connecting pieces, are connected by diagonal struts. The small beams of the framework are attached to the upper curvilinear strips of the chord connecting pieces in radial directions.

The flat radial and chord connecting pieces are made of welded tubular strips and I-shaped braces. The framework is made of class 3 steel. The cross section of the tubular strips of the connecting pieces is $\phi\,170 \times \delta_{pipe}$, where the wall thickness is $\delta_{pipe} = 10\text{-}30$ mm. The cross section of the tubular diagonal struts is $\phi\,170 \times 4$.

The upper curvilinear strips of the chord connecting pieces are connected: longitudinal steel strips and lateral ribs, to which the small beams of the framework are attached by four bolts and two studs, welded to the small beams at each end of the small beam, are welded to a $\phi\,170 \times 8$ mm pipe. The position of the small beam is regulated in relation to the framework by means of the studs. The small beams can be inserted into the framework in the intended position (for example, using an assembly control pattern) with an accuracy to 0.1-0.2 mm, which is checked using a test gage.

The covering of the reflecting surface of the reflector is attached to the small beams of the latticework by bolts or screws in threaded openings in the small beams. The curvilinear form of the small beams of the latticework for a particular parabola can be produced at the factory with a high accuracy of about ± 0.2 mm (for the upper rack) because it is a simple element. The shaping of the curvature of the small beams can be done on a large press with a press table equal in length to the length of the small beams, about 3000 mm. Since there are ten types of curvature for the small beams, the shaping of all the small beams for the latticework would require 10 dies. It should be noted that the shape of the die is made in the form of a parabolic cylinder into which a number of unshaped small beams are placed at the same time — 10 to 20 items, depending on the width of the press table. To be sure, the small beams can be shaped by other methods, such as by rolling, using shaping rollers, with individual adjustment using a control pattern.

In the variant considered, the reflector framework is sectional, with sections connected by class bolts. The connections to the chord connecting pieces and the diagonal struts are made in the form of two jaws (forks). Appropriate shapes are welded onto the strips of the radial supports. However, we feel that the framework should be fully welded. Welded connections of the supporting elements (radial supports, chord connecting pieces and diagonal struts) are made during erection work. In addition to lowering the cost of construction (there is no need for hundreds of small pieces and bolts), in a fully welded framework there is a considerable increase of its rigidity and there is full assurance of absence of gaps which may be present when bolting is used.

The sole inconvenience is possible buckling. But this inconvenience can be overcome because: a) possible buckling can be decreased by peening the welded seems, and, b) there is adequate leeway for adjusting the position of the small beams of the latticework, which is possible in the range ± 30 mm or more.

Unit of Small Reradiating Reflector and Four-Shaft Suspension Pyramid. The small reflector unit consists of a reradiating reflector 1.5-3 m in diameter and a counterweight attached to the vertex of a four-shaft pyramid using a design making it possible to change the arm of the counterweight (distance from the vertex of the pyramid).

The shafts of the pyramid are made in the form of thin-walled cylindrical casings 750 mm in diameter; wall thickness is $\delta_{wall} = 1.6$ mm, with diaphragms for increasing rigidity at intervals of 1200 mm. A 12-mm sheet of foam plastic is glued around the individual shafts for heat insulation from the direct rays of the sun.

The lower ends of the shafts of the pyramid are attached to the principal fulcrums of the reflector in the intermediate eight-support structure by means of ball-and-socket joints; the lengths of the shafts are adjustable.

The weights of the different units are given below:

Units	Wt., kg
Rotating sectors with axis and counterweight beam	516,000
Including concrete of counterweight	224,000
Intermediate eight-support structure	146,000
Octahedral shaft pyramid	4,700
Reflector	336,500
Including covering (sheet steel δ_{sh} = 2.2 mm)	52,300
Small reradiating reflector and four-shaft suspension pyramid	5,500
Total	1,315,000

Conclusions

1. The full solution of the problem of deformation of a parabolic reflector requires a more precise knowledge of the distribution of wind loads over the surface of the reflector for different angles of attack, and also a knowledge of the temperature gradients in the elements of the reflector under operating conditions.

2. A partial solution of this problem — constructing a reflector of considerable size with small deformations caused by the load from structural weight — can be attained by using a multisupport suspension of the reflector to the supporting-steering structure.

3. The principal characteristics of the design variant involving multisupport suspension of the reflector are: the supports have a radial-symmetric arrangement relative to the framework of the reflector (which also is constructed using a radial-symmetric supporting structure) and have the property of remaining in mutually parallel planes parallel to the aperture of the reflector, under the influence of both symmetric and skew-symmetric loads, as a result of the kinematic relationship to the turning of the reflector about its horizontal axis.

4. The advantage of this design is that it does not require complex apparatus for compensation of elastic deformations and therefore is reliable.

5. The described design variant of multisupport suspension of the reflector was used as the basis for preliminary design of a radio telescope with a 60-m reflector.

6. Use of this design for parabolic reflectors with a diameter 100 m or greater is possible when using servomechanisms ensuring compensation of elastic deformations by means of effect on one or two nodal points (such as the vertices of multishaft pyramids), and therefore operating with a high degree of reliability.

In conclusion, the author feels it a duty to express sincere appreciation to A. E. Salomonovich, senior scientific specialist at the laboratory, for many valuable comments on the content of the article, and for the attention which he devoted to the study, and also to V. T. Yevdokimova, I. A. Emel'yanov, V. P. Nazarov, and V. L. Shubeko, design engineers in the radio astronomy laboratory design office, for participating in designing the components described above and for work in finalizing this paper.

Literature Cited

1. P. D. Kalachev and A. E. Salomonovich, Trudy Fiz. Inst. Akad. Nauk SSSR, Vol. 17 (1962).
2. Yu. N. Pariiskii and S. É. Khaikin, Izv. Glavnaya Astrofiz. Observ. (Pulkovo), 21 (Issue 5, No. 164) (1960).
3. S. É. Khaikin, N. L. Kaidanovskii, N. A. Esepkina, and O. N. Shivris, Izv. Glavnaya Astrofiz. Observ. (Pulkovo), 21(Issue 5, No. 164) (1960).
4. A. Z. Fradkin, Antennas for Superhigh Frequencies, Izd. "Sovetskoe Radio" (1957), p. 406.

5.   G. Z. Aizenberg, Antennas for Ultrashort Waves, Svyaz'izdat (1957).

6.   J. Robieux, Ann. Radioélectr., 11(43) (1956).

7.   B. I. Klass, Aviation Week, 78(5): 80 (1963).

8.   L. Mohr, Stahlbau, 27(3) (1958).

9.   B. A. Garf, Use of Solar Energy, Vol. 1.

10. I. Feld, Ann. NY Acad. Sci., 70(2): 155-276.

11. I. G. Bolton, Radio Telescopes, Univ. of Chicago Press (1960).

12. V. V. Vitkevich, Transactions of the Fifth Conference of Problems in Cosmogony, Izd. Akad. Nauk SSSR (1956).

13. I. B. G. Hooghoubt, Schweiz. Techn. Z., 54(30/31): 673-676 (1957).

14. E. G. Bowen and H. C. Minnett, Proc. IRE Australia, 24(2) (1963).

15. Yu. L. Shakhbazyan, Izv. Glavnaya Astrofiz. Observ. (Pulkovo) (Issue 3, No. 172): 180-185 (1964).

RADIO LINES OF EXCITED HYDROGEN
AND THE FEASIBILITY OF DETECTING THEM*

R. L. Sorochenko

Hydrogen-line radiation generated by transitions $n-n'$ between atomic levels with different principal quantum numbers is described by the well-known Balmer formula

$$\nu = R \left(\frac{1}{n'^2} - \frac{1}{n^2} \right),$$

(1)

where R = 3.288057 · 10^{15} cps is the Rydberg constant and ν is the frequency of the radiation. In the optical range, a large number of such spectral lines, belonging to six series, have been observed experimentally. The sixth and last series, at the longest wavelengths, or rather its leading line (the transition from the seventh to the sixth level), at 12.37 μ, was only detected fairly recently, in 1953 [1]. Wild has pointed out the possibility that lines of excited hydrogen may be produced at radio wavelengths [2]. Kardashev subsequently calculated the intensities and widths of these lines for the wavelength range from 0.01 to 100 cm, and suggested that they might be observable in H II regions.

In these regions, electrons will be captured on upper levels upon recombination, and with subsequent cascade transitions a certain probability may be expected for emission of quanta with radio-wave energies. A very important factor in these processes is a "blurring" of the lines, caused mainly by interactions of excited hydrogen atoms with other particles — other atoms, electrons, and ions. Evidently the interaction effects will grow with wavelength, since transitions between strongly excited levels are responsible for the longer-wave radiation. It has been suggested [2] that, in the absence of any selective excitation mechanism, the discrete lines associated with transitions between levels having different principal quantum numbers should blend together in the radio range and merely supplement the continuous spectrum. On the other hand, from an approximate calculation for the pressure effect, Kardashev [3] concluded that the effect would be small for lines with frequencies above 7000 Mc, and that lines might be observable with existing equipment up to 1500 Mc.

Calculations are presented below for the emission intensity and line shapes in H II regions, as based on a recently developed theory for the interaction of excited atoms with the electric fields of electrons and ions.

Brightness Temperature in the Lines

To determine the intensity of the line emission, one must first evaluate the absorption coefficient corresponding to transitions $n-n'$ between levels of the hydrogen atom having large quantum numbers. According to [4], the absorption coefficient $k(\omega)$, allowing for stimulated emission, is given by

$$k(\omega) = \frac{1}{4}\, a_\omega \lambda^2 N_{n'} \frac{g}{g'} \left[1 - \frac{g'}{g} \frac{N_n}{N_{n'}} \right].$$

(2)

Here, N_n is the concentration of hydrogen atoms on the upper level n, $N_{n'}$ is the concentration on the lower level n', and the statistical weights g and g' of the upper and lower levels are equal to $2n^2$ and $2n'^2$, respectively. Also,

*Communicated to the Scientific Council of the Radio Astronomy Laboratory, April 27, 1963.

$$a_\omega = W^{SP}_{nn'} \cdot I(\omega), \tag{3}$$

where

$$W^{SP}_{nn'} = \frac{2\omega^2 e^2}{mc^3} f_{nn'} \tag{4}$$

is the Einstein probability coefficient for spontaneous transition from state n to state n'; $I(\omega)$ is the spectral density distribution over the line, normalized to unity; $f_{nn'}$ is the oscillator strength for the transition $n \to n'$; c is the velocity of light; e is the electron charge; m is the electron mass; $\omega = 2\pi\nu$ is the angular frequency; and λ is the wavelength.

Under conditions of thermodynamic equilibrium in an H II region of temperature T_e, the concentration N_n of hydrogen atoms in the $\bar{n}$-th quantum state is given [5] by the expression

$$N_n = b_n T_e N_i N_e \frac{h^3 n^2}{(2\pi mkT_e)^{3/2}} e^{-\frac{hR}{n^2 kT_e}}, \tag{5}$$

where N_i, N_e are the ion and electron concentrations; h is the Planck constant; and k is the Boltzmann constant.

For large n, as is the case for radio lines, we may take $b_n \approx 1$, and $\exp(hR/n^2 kT_e) \approx 1$. Since, under conditions of thermodynamic equilibrium *

$$\frac{g'}{g} \frac{N_n}{N_{n'}} = e^{-\frac{h\omega}{2\pi kT_e}}, \tag{6}$$

we have, upon substituting Eqs. (3)-(6) into Eq. (2)(with $h\omega \ll 2\pi kT_e$),

$$k(\omega) = \sqrt{\frac{\pi}{2}} \frac{e^2}{mc} \frac{h^3}{(mk)^{3/2}} \frac{h\omega}{2\pi kT_e} \frac{N_e^2 n^2}{T_e^{3/2}} f_{(n,n')} I(\omega). \tag{7}$$

For a transition $n - n'$ in a hydrogen atom, the oscillator strength is given [5] by the following expression, which neglects the Kramers–Gaunt factor:

$$f^+_{nn'} = \frac{2^5}{3\sqrt{3\pi}} \frac{1}{n^2} \frac{1}{\left(\frac{1}{n'^2} - \frac{1}{n}\right)^3} \left(\frac{1}{n^3} \frac{1}{n'^3}\right). \tag{8}$$

According to [6], the numerical coefficient in Eq. (8) is subject to an error of no more than a factor of two. This coefficient may be checked by passing to the classical case for large n. Let

$$f^+_{nn'} = \alpha \frac{1}{n^2} \frac{1}{\left(\frac{1}{n'^2} - \frac{1}{n^2}\right)^3} \frac{1}{n^3} \frac{1}{n'^3}. \tag{9}$$

Then

$$f^+_{n,n+s} = \frac{\alpha}{n^2} \frac{1}{\left[\frac{1}{n^2} - \frac{1}{(n+s)^2}\right]^3} \frac{1}{n^3} \frac{1}{(n+s)^3} \tag{10}$$

and

$$f^+_{n,n-s} = \frac{\alpha}{n^2} \frac{1}{\left[\frac{1}{n^2} - \frac{1}{(n-s)^2}\right]^3} \frac{1}{n^3} \frac{1}{(n-s)^3}. \tag{11}$$

---

*It is entirely possible that departures from thermodynamic equilibrium occur, since the cross sections of strongly excited states are not known. In view of the low densities, however, the distribution may evidently be considered a nearly Boltzmann one.

Adding Eqs. (10) and (11), and performing the transformation, we have

$$f_{n,n+s}^{+} + f_{n,n-s}^{+} = \alpha \, \frac{n^2}{s^2} \, \frac{24n^4 - 22n^2s^2 + 6s^4}{64n^6 - 48n^4s^2 + 12n^2s^4 - s^6} \, . \tag{12}$$

Since the sum of the oscillator strengths over all transitions is

$$\sum_{s=1}^{n} (f_{n,n+s}^{+} + f_{n,n-s}^{+}) = 1, \tag{13}$$

we may evaluate α by substituting Eq. (12) into Eq. (13). A calculation performed on an electronic computer yielded the value $\alpha = 1.62$ for large n. With this coefficient for the transition between adjacent orbits, n = n' + 1, we have $f_{nn'}^{+} = 0.2$ n, or, including a Kramers–Gaunt factor, $\gamma = 0.75$ [3],

$$f_{nn'} = \gamma f_{nn'}^{+} = 0.15 \, n. \tag{14}$$

Introducing Eq. (14) and the numerical values of the constants into Eq. (7), we obtain

$$k(\omega) = \frac{3.3 \cdot 10^{-12} N_e^2}{T_e^{5/2}} \, I(\omega). \tag{15}$$

In deriving Eq. (15), it has been assumed that for large n the Balmer formula (1) will yield $n^3 \nu \simeq 2R$. Having the absorption coefficient, we can readily determine the optical depth τ and the brightness temperature $T_{b.1}$ of the line; for $T_e = 10^4$ °K, we have

$$\tau(\omega) = k(\omega) \, l = 10^{-3} \, ME \, I(\omega); \tag{16}$$

$$T_{b.1} = T_e \tau(\omega) = 10 \, ME \, I(\omega). \tag{17}$$

Here, $ME = N_e^2 l$ is the emission measure, with dimensions of cm^{-6} pc, and l is the path length.

Line Widths and Profiles

The width of an emission line is governed by radiation losses, Doppler broadening, and the interaction of the radiating atom with other particles. In the great majority of cases, including the case considered here, the last two factors are the relevant ones.

For the case of pure Doppler broadening, the spectral density distribution is given by the expression

$$I(\omega) = \frac{1}{\Delta\omega_{Dop}\sqrt{\pi}} \, e^{-\left(\frac{\omega - \omega_0}{\Delta\omega_{Dop}}\right)^2}, \tag{18}$$

where

$$\Delta\omega_{Dop} = \frac{\omega_0}{c} \sqrt{\frac{2kT_e}{m_H}} \tag{19}$$

is the halfwidth of the line for a fall in intensity by a factor of e, m_H is the mass of the hydrogen atom, and ω_0 is the angular frequency at the line center. According to Eqs. (17) and (18), the brightness temperature at the line center is

$$T_{b.l.c.}^{Dop} = \frac{5.65 \, ME}{\Delta\omega_{Dop}} = 0.9\lambda ME \sqrt{\frac{m}{2kT_e}} \, . \tag{20}$$

The broadening of excited hydrogen lines through interaction of atoms with electrons and ions has been treated thoroughly in [7]. In this theoretical treatment, the first to take rigorously into account the influence of the electric fields of electrons and ions, it is shown that even for relatively low electron densities the electric fields can induce a rather considerable line broadening. The electric fields of the electrons, which had not previously been considered with reference to large quantum numbers, are found to make the predominant contribution. According to these results [7], the line profile is described by the so-called dispersion formula

$$I\,(\omega) = \frac{1}{\pi} \frac{\Delta\omega_{disp}}{(\Delta\omega_{disp})^2 + (\omega - \omega_0)^2}, \tag{21}$$

and the halfwidth to the 0.5 level in the line is given by the expression

$$\Delta\omega_{disp} = \frac{1}{9} \left(\frac{8\pi m}{kT_e}\right)^{1/2} N_e \left(\frac{h}{m}\right)^2 \int\limits_{y_{min}}^{\infty} \frac{e^{-y}}{y}\, dy\, (n^5 + n'^5), \tag{22}$$

where

$$y_{min} = \frac{4\pi N_e}{3m} \left(\frac{ehn^2}{kT_e}\right)^2.$$

For the case n' = n − 1 ≈ n, straightforward transformation of Eq. (22) yields the following simple formula for further calculation:

$$\Delta\omega_{disp} = 1.75\cdot 10^{-5} N_e T_e'^{/2} \left[\log \frac{4\cdot 10^6 T_e}{n^2 N_e^{1/2}} - 0.125\right] n^5. \tag{23}$$

The brightness temperature at the line center for the dispersion profile is equal, by Eqs. (17) and (21), to

$$T_{b.l.c.}^{disp} = \frac{5.65 ME}{\sqrt{\pi}\, \Delta\omega_{disp}}. \tag{24}$$

Comparing the expressions (20) and (24), we note the appearance of the factor $\sqrt{\pi}$ in the denominator of (24) because of the extended wings of the dispersion profile; as a result, the spectral density at the line center is considerably reduced.

Under the conditions prevailing in HII regions, the width and shape of the line profile will be established both by Doppler broadening and by the interaction effect. In this event, the profile shape will be described by the expression

$$I\,(\omega) = \frac{a}{\pi^{3/2}} \int\limits_{-\infty}^{\infty} \frac{\Delta\omega_{disp}\, e^{-a^2 u^2} du}{\Delta\omega_{disp}^2 + \left(\omega - \omega_0 + \frac{u}{c}\,\omega\right)^2}, \tag{25}$$

where $a = \sqrt{m_H/2kT_e}$.

Equation (25) may be transformed in the following way:

$$I\,(\omega) = \frac{ac}{\omega\,\sqrt{\pi}} \frac{1}{\pi} \int\limits_{-\infty}^{\infty} \frac{y e^{-t^2}}{y^2 + (x + t)^2}\, dt, \tag{26}$$

where $y = \Delta\omega_{disp} ac/\omega$ and $x = y(\omega - \omega_0)/\Delta\omega_{disp}$. Since the quantity

$$z\,(x, y) = \frac{1}{\pi} \int\limits_{-\infty}^{\infty} \frac{y e^{-t^2}\, dt}{y^2 + (x + t)^2} \tag{27}$$

is a probability integral, the spectral density distribution (26) in the line can be obtained from tabulated values of the integral [8]. With Eqs. (17), (26), and (27), it is evident that the brightness temperature at the line center for the case of combined Doppler broadening and interaction effects is given by the expression

$$T_{b.l.c.} = \frac{1.32\cdot 10^5 ME}{\omega}\, z\,(0, y), \tag{28}$$

where z(0,y) is the probability integral for x = 0.

Experimental Possibilities of Observing Lines in HII Regions. As Eqs. (20), (24), and (28) indicate, the maximum line intensity should be observed in emission nebulae having a large emission

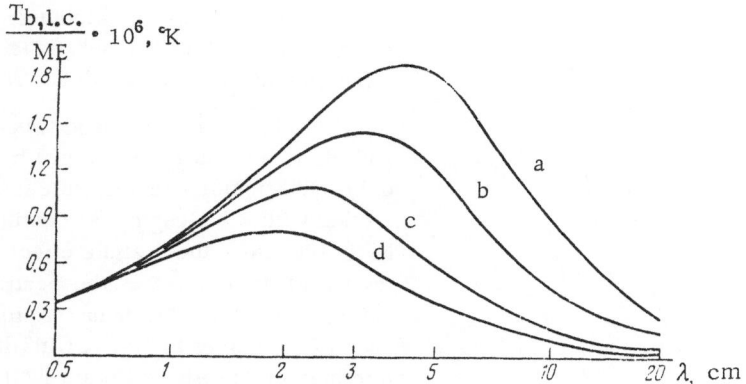

Fig. 1. Brightness temperature at the center of a line as a function of wavelength for selected electron concentrations in an H II region. a) $N_e = 100$ cm^{-3}; b) $N_e = 200$; c) $N_e = 500$; d) $N_e = 1000$.

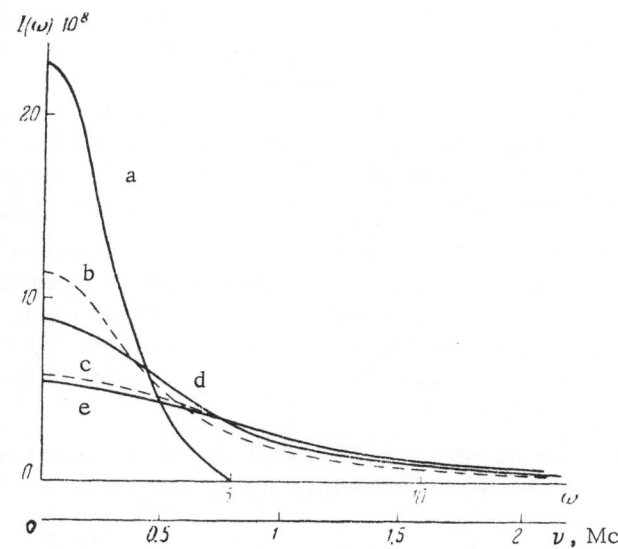

Fig. 2. Profile of a spectral line, $\lambda = 3$ cm, computed for five cases. a) Pure Doppler broadening; b,c) broadening produced by interaction of excited atoms with electrons and ions for $N_e = 500$ and 1000 cm^{-3}, respectively; d,e) combined Doppler and interaction broadening for $N_e = 500$ and 1000 cm^{-3}, respectively. Abscissa, angular and cyclic frequency; ordinate, spectral density distribution normalized to unity.

measure. On the other hand, according to Eqs. (23) and (24), it is important that the emission measure ME = $N_e^2 l$ not rest on large electron densities but on large nebular dimensions, since at low densities the interaction of excited atoms with electrons and ions has a weaker effect. According to the catalogs [9, 10], the brightest H II regions, with ME $\approx 10^6$, are the emission nebulae NGC 1976 (Orion) and NGC 6618 (Omega). The next group of nebulae, NGC 6523 (Lagoon), IC 1795, and Sagittarius A, have ME $\approx 10^5$. The mean electron concentrations in these nebulae are $\approx 10^3$ cm^{-3} for the Orion, 500 for the Omega, 200 for the Lagoon, and 100 for the last two.

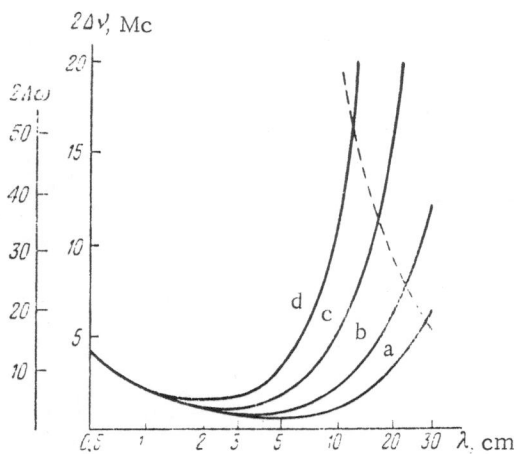

Fig. 3. Line width as a function of wavelength and electron concentration. a) $N_e = 100$ cm^{-3}; b) $N_e = 200$; c) $N_e = 500$; d) $N_e = 1000$. Broken curve, limit of detectability of lines against background continuum.

Figure 1 shows the brightness temperature at the center of a line (normalized to the emission measure) as a function of the wavelength and N_e, according to Eq.(28).

In the millimeter range, where $\Delta\omega_{Dop} \gg \Delta\omega_{disp}$ and the line width is determined by Doppler broadening only, the brightness temperature (20) of the lines increases with wavelength. At centimeter wavelengths for $N_e > 100$ cm^{-3}, the pressure effect begins to enter, and as the wavelength increases, the spectral density of the lines declines sharply. This behavior produces the turnover in each of the curves in Fig. 1, with the brightness-temperature maximum shifting toward shorter wavelengths for large densities. The latter circumstance is readily understood from a physical standpoint, since the longer-wave radiation is generated by atoms whose electrons are in larger orbits. Such atoms will naturally be more likely to interact with electrons and ions, resulting in a "smearing" of the lines.

Figure 2 presents line profiles computed from Eqs. (18), (21), and (26) for $N_e = 1000$ and $N_e = 500$. A wavelength of 3 cm is taken as an example. As the figure shows, the pressure effect has already reduced the spectral density at the line center by a factor of three for an electron concentration of 500, and by a factor of six for a concentration of 1000. As the wavelength increases, the interaction of excited atoms with electrons and ions will lead to an even sharper decline in spectral line density as well as a broadening, ultimately resulting in a loss of line contrast against the background continuum.

Figure 3 illustrates how the line width depends on wavelength and electron concentration, according to Eq. (26). We see that for the concentrations indicated the pressure effect, which causes an increase in line width with wavelength, begins to appear at wavelengths near 1 cm. Moreover, the effect is stronger for higher concentrations, as follows from the physics of the process. The broken curve in the figure corresponds to the wavelengths where the line width is one-third of the distance between adjacent lines, and in a sense represents a boundary criterion for distinguishing the lines.

Since atomic-hydrogen lines are formed in H II regions, the line emission will be superposed on the radiation of the continuous spectrum. The absorption coefficient for a diffuse, ionized interstellar medium is given [11] by

$$k(\omega) = \frac{16\,\pi^2}{3\,\sqrt{2\pi}}\,\frac{e^6}{(mkT_e)^{3/2}}\,\frac{1}{c\omega^2}\,N_e N_i \ln\frac{(2kT_e)^3}{4.6\omega^2 e^4 m}\,, \tag{29}$$

while the brightness temperature of the continuum radiation is

$$T_H = \frac{5.77\cdot 10^{17}}{T_e^{1/2}\,\omega^2}\ln A, \tag{30}$$

where

$$\ln A = \ln\frac{(2kT_e)^3}{4.6\,\omega^2 e^4 m}\,.$$

By comparing Eqs. (28) and (30), we obtain the relative brightening of an H II region in the lines:

$$\frac{T_{b.l.c.}}{T_H} = \frac{0.23\cdot 10^{-10}\,\omega}{\ln A}\,z(0,y). \tag{31}$$

A graphical representation of this quantity as a function of wavelength is given in Fig. 4. Since the brightness temperature in the lines is proportional to the first power of the wavelength [Eq. (20)], while the

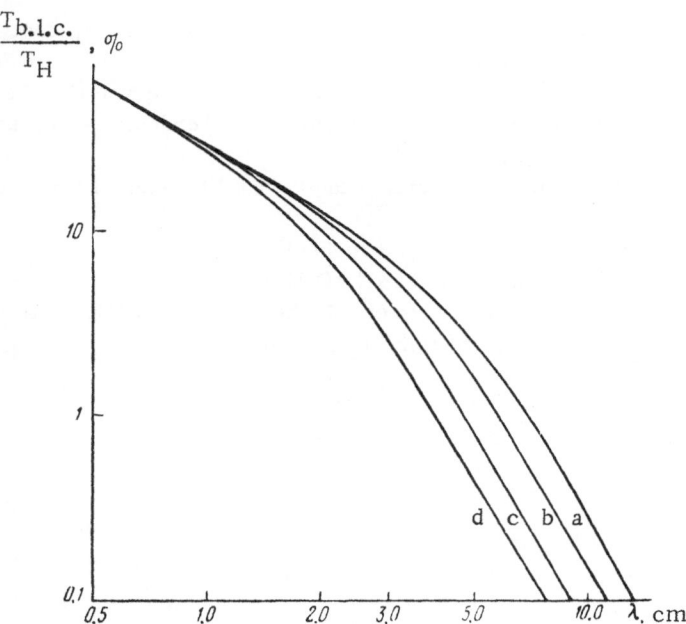

Fig. 4. Relative enhancement of line brightness in an H II region as a function of wavelength and electron concentration. a) N_e = 100 cm^{-3}; b) N_e = 200; c) N_e = 500; d) N_e = 1000.

brightness in the continuum increases more rapidly than λ^2, the line contrast will decrease with wavelength even at wavelengths where the pressure has not yet entered. Pressure effects will reinforce this decline and lead to a virtually complete "smearing-out" of the lines in the 10- to 15-cm region.

From the viewpoint of an experimental attempt to detect the lines, the decline in the line brightness temperature with wavelength because of the pressure effect may partially be compensated by the opportunity to expand the bandwidth of the radiometer, which would increase its fluctuation sensitivity δT. Thus, the maxima in the $T_{b.l.c.}/\delta T$ curves will be somewhat further displaced toward longer wavelengths than the maxima in the $T_{b.l.c.}$ curves shown in Fig. 1.

For radio telescopes 22-25 m in diameter, the antenna temperatures in the lines (allowing for dispersion and smoothing by the antenna beam) should, in the optimum wavelength range of 1.5-4 cm, amount to a few tenths of a degree Kelvin for the brightest nebulae. According to the considerations given above, the maximum line brightness at radio wavelengths would be expected at 2.5-3 cm in the direction of the Omega nebula, which together with the Orion nebula has the greatest emission measure, although a somewhat lower density.

To record such small antenna temperatures with a relatively narrow ($\approx$ 1 Mc) receiver bandwidth and a weak contrast against the background continuum of the source would be a fairly difficult task, although it would be within the capability of current techniques of radio-astronomical measurement. Apart from a high-sensitivity spectrometer, the equipment used should be nonsusceptible to fluctuations in the continuum intensity, which will inevitably occur during the observing period, both because of inaccurate guiding on the source and changes in the antenna temperature arising from radiation by the earth and the atmosphere.

The author considers it a pleasant obligation to express particular appreciation to I. I. Sobel'man for valuable counsel and discussions while the work was in progress.

Literature Cited

1. C. Hamphreys, J. Res. Bur. St., 50:1 (1953).
2. I. Wild, Astrophys. J., 115:206 (1952).
3. N. S. Kardashev, Astr. Zh., 36:838 (1959).

4. I. I. Sobel'man, Introduction to the Theory of Atomic Spectra, Moscow, Phys. Math. Press (1963).

5. D. Menzel, Astrophys. J., 85: 330 (1937).

6. H. Bethe and E. Salpeter, Quantum Mechanics of Atoms with One and Two Electrons [Russian translation], Moscow, Phys. Math. Press (1960). Published in English by Academic Press, Inc., New York.

7. H. R. Griem, Astrophys. J., 132: 883 (1960).

8. V. N. Fadeeva and N. M. Terent'ev, Tables of the Probability Integral for a Complex Argument, Moscow, Techn. Theor. Press (1954).

9. A. D. Kuz'min, Tr. Fiz. Inst. Akad. Nauk SSSR, 17: 84 (1962).

10. G. Westerhout, Bull. Astron. Inst. Neth., 14: 215 (1958).

11. I. S. Shklovskii, Cosmic Radio Waves, Moscow, Techn. Theor. Press (1956) [English translation by R. B. Rodman and C. M. Varsavsky, Harvard Univ. Press (1960)].

STATISTICAL ESTIMATION OF THE EFFECT OF THE PRECISION
AND RIGIDITY OF A RADIO-TELESCOPE ANTENNA
ON ITS PARAMETERS

A. E. Salomonovich

The parameters of the antenna of a radio telescope (the directional diagram, the effective area, etc.), are influenced by inaccuracies in the antenna construction and by various types of deformations. * The effect of random errors in the preparation of the antenna surface on its directional diagram, gain, and effective area has been discussed by several authors [2-5]. In particular, Shifrin [5] has derived formulas relating the dispersion and correlation radius for random phase errors in the aperture of a line-source antenna, as caused by inaccurate construction, with deformation of the beam and deterioration of the gain. Shifrin has also considered the analogous problem for aperture systems [6]. The developments in these and certain other papers should, however, be supplemented by considerations that apparently have not yet been taken into account. There are, in fact, at least two different effects that might be responsible for distorting the surface of an actual antenna, thereby resulting in phase distortions of the field.

An antenna surface can only be built to a certain finite precision, characterized by the value of the maximum relative tolerance $\varepsilon_p/L = 10^{-m_p}$ (where L is the diameter of the antenna aperture), the controlling phase dispersion σ_p, and the correlation radius ρ_p. These parameters can be measured after the antenna has been built, and the measurements usually apply for the specific external conditions under which they were obtained.

When an antenna is placed in service, it is subject to several disturbances — it experiences weight, wind, and thermal deformations. Its surface is thereby changed. These changes induce phase distortions in the field, leading to a further deterioration in the antenna characteristics. The distortions caused by deformation differ from the distortions caused by imprecise construction, in that the former are always more even — sections that are deformed simultaneously are commensurable with the antenna aperture. Moreover, the deformations ordinarily do not remain invariant.

In the great majority of cases we have no opportunity to make a detailed study of the character of the phase distortions by deformation, since they depend in a complicated way on the antenna design and do not remain constant. Thus, the phase distortions due to deformation of an antenna of given aperture cannot be described analytically, such as by development in power series. In fact, it is expedient to regard these distortions as random, like the distortions due to imprecise construction. We may consider here an ensemble of possible configurations for the deformed antenna surface, or a configuration depending randomly on time. Just as when treating the effects of imprecision, for small deformations the mean antenna parameters will differ little from the parameters representing an individual situation. In this approach, we would regard the antenna surface as deformed in a random manner, but with prescribed characteristics: a maximum deformation $\varepsilon_d/L = 10^{-m_d}$, a fixed dispersion σ_d, and a correlation radius ρ_d.

The statistical characteristics of the distribution of phase errors due to imprecise construction and to deformations will, as a rule, differ substantially from one another; usually $\rho_p \ll \rho_d$. For the quantities σ_p and σ_d,

---

* The effects of scattering by the feed supports and other structural elements are considered in the following paper [1].

widely varying relationships are possible. A precise by insufficiently rigid reflector will have $\sigma_d > \sigma_p$, while for a coarsely built but rigid reflector, $\sigma_d < \sigma_p$. In general, the deformation phase distortions will depend on time. Nevertheless, over fairly long intervals of time the antenna might be regarded as having a surface of constant shape which we do not know exactly, but have to characterize statistically.

Since the two effects considered are independent, for small distortions we can determine the value of the true gain D of a given antenna by the expression

$$\overline{D} = D_0 \, (1 - \triangle_p) \, (1 - \triangle_d), \tag{1}$$

where D_0 is the gain of an ideal antenna,* and $\triangle_p$, $\triangle_d$ are the relative losses in gain resulting from the known quantities σ_p, ρ_p and σ_d, ρ_d, respectively. For definiteness we shall suppose that the imprecision distortions are distributed approximately normally, so that $\varepsilon_p \approx 2.6\lambda/4\pi\sigma_p$, where $\sigma_p^2 \approx 25(\varepsilon_p/\lambda)^2$. Moreover, we will usually have $\rho_p = n\lambda$, where for pencil-beam antennas n = 1-20.

To estimate the order of magnitude of σ_d and ρ_d, we may regard the deformation displacements (along normals) of the surface points from a prescribed profile as described by the equation of an elastic line. A calculation shows that for approximate estimates we may use $\varepsilon(x) = \varepsilon_d x^2$, where x = 2r/L, and r is the distance from the center of the circular aperture. Excluding constant displacements along the surface, we obtain for the effective mean value of the dispersion:

$$\sigma_d^2 = \overline{\varphi_d^2(x)} = \int_0^1 \varphi_d^2(x) \, dx \simeq 14 \, (\varepsilon_d/\lambda)^2. \tag{2}$$

The phase distortion $\varphi_d(x)$ changes sign for $x = 1/\sqrt{3} \approx 0.6$. We shall adopt the corresponding approximation $\rho_d \approx 0.3L$. For the case of pencil-beam antennas, as pointed out above, $\rho_p/\lambda = 1-20$; moreover, $\lambda/L = 10^{-3}-10^{-4}$. With these values for the parameters we obtain, by [5],

$$\triangle_p \approx \sigma_p^2 = 25 \, (\varepsilon_p/\lambda)^2. \tag{3}$$

On the other hand, for $\rho_d/\lambda \geq 1/\pi$, the quantity $\triangle_d = F\sigma_d^2$, as shown in [5], where F is an almost linear function ranging from 1 as $\rho_d/L \to 0$ to 0.4 for $\rho_d/L = 0.75$.

If we express σ_p and σ_d in terms of the parameters m_p and m_d introduced above, and retain only first-order terms in σ_p^2 and σ_d^2, we will obtain

$$D = 10h^2 \, [1 - 0.25 \, (10^{2\,(1-m_p)} + 0.6F \cdot 10^{2\,(1-m_d)}) \, h^2]. \tag{4}$$

The limiting value of the gain that can be achieved for prescribed tolerances on the precision and rigidity is found to be

$$D_{\max} = 10^{2m_p-1}/(1 + 0.6F \cdot 10^{2\,(m_p-m_d)}, \tag{5}$$

or in decibels

$$N_{\max} = 20m_p - 10 - 10 \log \, [1 + 0.6F \cdot 10^{2\,(m_p-m_d)}. \tag{6}$$

If the tolerance on structural precision is considerably greater than that permitted for the rigidity of the reflector, i.e., if $m_d \gg m_p$, then

$$N_{\max} = 20m_p - 10, \tag{7}$$

---

* To simplify the computations we shall take $D_0 = \pi^2h^2 \simeq 10h^2$, where h = L/$\lambda$, i.e., we regard the aperture utilization factor as unity.

a value readily seen to correspond to that obtained without allowance for deformations [2, 5]. If, on the contrary, the reflector is constructed relatively precisely but does not have much rigidity, i.e., if $m_d \ll m_p$, then [in the event that $10^{2(m_p-m_d)} \gg 1$, as is usually the case],

$$N_{max} \approx 20m_d - 10 - 10 \log [0.6F], \tag{8}$$

so that the gain will be limited by the insufficient rigidity of the antenna. In this case, N_{max} is found to be somewhat greater than in Eq. (7). This is because of the relatively smaller effect of the deformation distortions, which have a larger correlation radius. For example, for $\rho_d = 0.3L$, the quantity $N_{max} = 20m_d - 5$. If $m_p = m_d$,

$$N_{max} \approx 20m_p - 11, \tag{9}$$

and there will be only a minor additional deterioration in the limiting gain because of the effect of insufficient rigidity.

The difference in the correlation radii ρ_p, ρ_d of the phase distortions enables one to estimate the factors responsible for deterioration of the antenna characteristics in comparison with ideal values by measuring them. The loss in gain because of imprecision, with $\rho_p \ll L$, leads to an increase in isotropic scattering although there is no change in the shape of the main lobe. But the loss in gain because of deformations, where $\rho_d \approx L$, results largely in an expansion of the main lobe.

By measuring the width of the main lobe of a pencil-beam antenna,* one may use the measured expansion for known ρ_d to estimate the value of σ_d, and thereby Δ_d. A measurement of the total loss in gain then would enable one to isolate the portion caused by imprecise construction, i.e., Δ_p. The latter quantity can be determined independently by measuring the scattering coefficient of the antenna. By observing the dependence of the width of the main lobe on external conditions one can also obtain information on the character of the deformations responsible for expansion of the lobe (weight deformations, thermal deformations, and the like).

We have applied the considerations set forth above to investigate the properties of the parabolic pencil-beam antenna of the RT-22 radio telescope. A survey of the reflector surface of the RT-22 showed that $\rho_p \approx$ 200 mm, and that $\sigma_p^2 = 0.21$ for $\lambda = 8$ mm. A computation yields $D/D_0 = 0.81$ for these conditions. Measurement of the width of the main lobe indicated that the relative expansion amounts to about 6%. It follows from an analysis of elastic deformations that the expected $\rho_d \approx 0.3L$. For this value of ρ_d, the measured expansion of the main lobe corresponds to $\sigma_d^2 = 0.30$. If the deformations increase quadratically from the center to the edge of the reflector, one obtains a mean value $\varepsilon_d = 1.3$ mm, in close agreement with the value obtained computationally [8]. Since we know σ_d^2 and ρ_d, we can estimate Δ_d; the value is found to be $\Delta_d = 0.17$. If we furthermore take into account the influence of nonuniform irradiation of the aperture and scattering by the feed supports and other structural elements [1], we can estimate the total loss in gain. As a result, the effective antenna area of the RT-22 radio telescope at wavelength $\lambda = 8$ mm was found to be close to 100 m^2, corresponding to about 26%, of the geometric area of the aperture. This computed value agrees with the results of measurements of the effective area [7].

It will be evident, then, that in cases where one cannot make an exact calculation for the expected deformations, the statistical method of estimating the combined phase distortions is suitable both for formulating requirements for the design of large reflecting radio-telescope antennas, and for estimating the values to be expected for the antenna parameters.

Literature Cited

1. P. D. Kalachev and A. E. Salomonovich, this volume, p. 7.
2. J. Robieux, Ann. Radioelectr., 11(43): 29 (1956).
3. O. N. Vladimirova, Dissertation, Moscow Phys. Techn. Inst. (1959).

---

* The measuring technique is described in [7].

4.   B. V. Braude, N. A. Esepkina, N. L. Kaidanovskii, and S. É. Khaikin, Radiotekhn. i Elektron., 5:584 (1960).
5.   Ya. S. Shifrin, Statistics of the Field of a Line-Source Antenna (1962).
6.   Ya. S. Shifrin, Akust. Zh., 7:248 (1961).
7.   A. E. Salomonovich, B. V. Braude, and N. A. Esepkina, this volume, p. 116.
8.   P. D. Kalachev, Tr. Fiz. Inst. Akad. Nauk SSSR, this volume, p. 143 and p. 161.

HOW THE EFFECTIVE ANTENNA AREA OF A RADIO TELESCOPE CAN BE INCREASED BY DECREASING SCATTERING ON ANTENNA BRACING STRUCTURES

P. D. Kalachev and A. E. Salomonovich

The parameters of a radio telescope reflector antenna, and its effective area in particular, are affected by inaccuracies in fabrication and by various deformations of the surface. The field established by the radiation over the antenna aperture is further distorted by structures supporting the feed, by bracing bars, and by the rim of the reflector, all of which act as scatterers. The effect of scattering on the rim and on bracing structures on the effective antenna area has been discussed in [1] and elsewhere. Calculations are reported only for the infrequent case where the bracing bars bear on the reflector rim.

The actual design features of the reflector must be taken into account in calculating scattering on large reflector antennas of radio telescopes. Accurate treatment of these design features can help in reducing loss of effective antenna aperture in radio telescopes.

This article takes up scattering on the bracing bars of a large reflector radio telescope and ways of coping with it.

Scattering on Bracing Bars and Effective Antenna Area of a Radio Telescope

We shall not be interested here in the field distribution in side lobes and back lobes due to scattering on braces situated in front of the reflector aperture or on the rim of the reflector itself; rather, our interest is limited to the energy loss from the principal lobe which is responsible for the drop in directive gain in the main direction. We shall, therefore, use the method presented by Kinber [1], who derived formulas for the relative power scattered on the rim of a reflector antenna and on its bracing structures.

The field distribution in the aperture is assumed to decline by the law

$$\frac{E(r)}{E_0} = 1 - (1 - q)(r/R_0)^2, \tag{1}$$

where R_0 is the reflector aperture radius, and q is the level of the field amplitude and the reflector rim. When the power level at the rim is −10 dB relative to the center, q = 0.32.

The total power flux radiated by the aperture at a field distribution (1) is obtained by integrating (1) over the aperture

$$w_p = \frac{E_0^2 \pi R_0^2}{2\pi}\left[q - \frac{1}{3}(1 - q)^2\right], \tag{2}$$

where Ω = 120 ohm is the total free-space impedance. As calculations show [1], the power escaping into side lobes by scattering from the rim is

$$w_{rim} = 2\pi R_0 \frac{E_0^2}{2\Omega}\sigma_{rim}, \tag{3}$$

77

where σ_{rim} is the linear scattering cross section of the rim. Our calculations of σ_{rim} reduce to the relative power loss by scattering on the rim

$$\varkappa_1 = \frac{w_{rim}}{w_p} = \frac{2}{\pi\left[q + \dfrac{1}{3}(1-q)^2\right]}\left\{\frac{1}{\mu}\left[q^2 + \frac{2}{3}\left(\frac{f \arccos P}{R_{rim}\mu}\right)\right] + \frac{1}{2\pi}\sqrt{\frac{R_{rim}\lambda}{R_0^2}}\right\}. \tag{4}$$

Here, for a paraboloid of rotation, $\mu(q) = 5.2 - 1.3q$; $R_{rim}/R_0^2 = f/R_0^2 + 1/4f$, where f is the focal length, $P = qR_{rim}/f$, R_{rim} is the distance from the focus to the reflector rim.

The problem of finding the power fraction scattered from braces supported on the reflector rim actually reduces in [1] to the familiar problem of scattering of a plane wave incident on a cylindrical obstacle. For the specified field distribution pattern over the aperture (1), the relative power scattered by the structures is

$$\varkappa_2 = \frac{w_T}{w_p} = \frac{\left[\dfrac{1}{3}(1+2q) + \dfrac{1}{5}(1-q)^2\right]}{\pi r_0\left[q + \dfrac{1}{3}(1-q)^2\right]}\sigma_{\perp,\parallel}, \tag{5}$$

where $\sigma_{\perp,\parallel}(2\pi r_0/\lambda)$ is the scattering cross section of the cylinder per unit length for either cross polarization ($\perp$) or longitudinal polarization ($\parallel$), and r_0 is the cylinder radius.

As demonstrated in the article by Potekhin [2], when a plane wave is incident on a cylinder of radius r_0, the linear scattering cross section is $\sigma_{\perp,\parallel}[(2\pi/\lambda)r_0\sin\psi]$, where ψ is the angle formed by the cylinder axis with the normal to the front of the incident wave. $\sigma_{\parallel}$ and $\sigma_{\perp}$ were determined by Borgnis and Papas [3]. When $2\pi/\lambda r_0\sin\psi \gtrsim 10$, the values of $\sigma_{\parallel}$ and $\sigma_{\perp}$ differ only slightly from $4r_0\sin\psi$, i.e., they are close to double the projection of the cylinder's geometric shadow. In the geometric optics approximation, the incident wave loses not only the power incident on the cylinder, but also an equal amount of power associated with the field, whose interference with the arriving wave sets up a zero field in the geometric shadow region. This inference extends to the case of scattering on obstacles of any configuration. In particular, in order to estimate the fraction of power scattered by the antenna feed, we can use the formula valid when $2\pi c/\lambda \gg 1$, where c is the radius of a disk equivalent to the feed structure

$$\varkappa_3 = \frac{w_{sh}}{w_p} = \frac{2c^2}{R_0^2\left[q + \dfrac{1}{3}(1-q)^2\right]}. \tag{6}$$

Adding up $\varkappa_1$, $\varkappa_2$, and $\varkappa_3$, we obtain the total relative power escaping to the side lobes through scattering on rim, braces, and feed.

The assumption up to this point has been that the braces bear on the reflector rim. But this situation is not the usual one. When strict requirements for maintaining an exact surface are imposed on a reflector antenna of fairly large size, i.e., the reflector is required to resist deformation, additional deformations of the reflector could result, generally speaking, from having the bracing bars supported on the rim, and this could impair reflector performance. For that reason, the braces (bars) bearing the load of the feed and the high-frequency equipment in large radio telescopes are usually supported on an intermediate force-bearing chord situated not on the circle of radius R_0, but on a circle of lesser radius (b) (cf. Fig. 1).

This situation obtains, in particular, on the high-precision RT-22 radio telescope [4, 5], since the problem of maintaining surface exactness required that the reflector be relieved of the reactions exerted by bracing bars supported on the force-bearing elements situated on the circumference of a circle of radius 5.5 m,

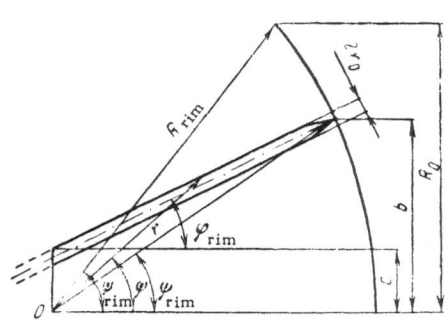

Fig. 1. Illustration for calculations of shadowing geometry.

where the reflector aperture radius was 11 m. This arrangement of bracing bars involved, as we shall see, an increased scattering of the radiation picked up by the radio telescope.

Consider, for definiteness, the case of four symmetrically placed bracing bars (Fig. 1). A part of the wavefront inside a cone of vertex angle $2\psi_T$ is scattered after reflection from the reflector as a plane wave. But the peripheral portion of the spherical wavefront in the solid angle bounded by cones of vertex angles $2\psi_{rim}$ and $2\psi_T$ is scattered by bracing bars before the reflection as a spherical wave.

The power associated with a plane wave scattered by the bracing bars can be shown to be

$$\omega_{T_1} = \frac{E_0^2}{2\Omega} R_0 N \sigma_{bar} I_1,$$

(7)

where N is the number of bracing bars, and σ_{bar} is the linear scattering cross section of a cylindrical bracing bar. The integral I_1

$$I_1 = \frac{b-c}{R_0} - \frac{2}{3}(1-q)\frac{b^3-c^3}{R_0^3} + \frac{1}{5}(1-q)\frac{b^5-c^5}{R_0^5},$$

(8)

where c is the distance from the focus to the point where the bar intersects the focal plane. The power associated with a spherical wave scattered by bracing bars may be estimated as follows. We compute the power radiated by the part of the aperture plane taken over the geometric shadow region formed by the spherical wave incident on the cylindrical bracing bar, and then assume that for all large values of the parameter $(2\pi r_0/\lambda)\sin\psi$ (r_0 being the radius of the bracing bar) the scattering cross section will be double the area of the geometric shadow, just as in the solution of the plane wave scattering problem [3].

For the case where the parameter $(2\pi r_0/\lambda)\sin\psi$ cannot be assumed large, we may resort to the averages of $\sigma_\perp$ and $\sigma_\parallel$. Assuming as above the field distribution (1), we obtain the power scattered by the peripheral portion of the spherical wave:

$$w_{T_2} = \frac{2E_0^2}{\Omega} N r_0 (I_1 + I_2),$$

where

$$\left.\begin{array}{l} I_1 = (R_0 - b) - \dfrac{2}{3}\dfrac{(1-q)}{R_0^2}(R_0^3 - b^3) + \dfrac{(1-q)^2}{5R_0^4}(R_0^5 - b^5); \\[2mm] I_2 = \dfrac{(R_0-b)^2}{2} - \dfrac{2(1-q)}{R_0^2}\left[\dfrac{R_0^4 - b^4}{4} - \dfrac{b(R_0^3 - b^3)}{3}\right] + \dfrac{(1-q)}{R_0^4} \times \\[4mm] \qquad\qquad \times \left[\dfrac{R_0^6 - b^6}{6} - \dfrac{b}{5}(R_0^5 - b^5)\right]. \end{array}\right\}$$

(9)

Another variant is as follows. The distribution of the field strength of the spherical wave from the feed in the direction of an obliquely placed bar (Fig. 1) appears in the form

$$E(\psi) = \frac{fE_0}{c}\left(\sin\psi - \frac{\tan\varphi_T}{\cos\psi}\right)\cos^n\psi.$$

(10)

Using the conditions for radiation at the reflector rim, we may find the power exponent $n = \log[cR_{rim}/f] \cdot \log\cos\psi_{rim}$, and the field distribution pattern of the incident wave along the rod. We have

$$w_{T_2} = \frac{f^2 E_0^2}{2c\Omega} N \int_{\psi_T}^{\psi_{rim}} \sigma\left[\frac{2\pi}{\lambda} r_0 \sin(\psi - \varphi_T)\right][\cos^{2n}\psi \sin\psi - \tan\varphi_T \cos^{2n-1}\psi]\,d\psi.$$

(11)

Assuming $\sigma = \text{const} = 4r_0$, i.e., the same as for a plane wave, we get, after integrating,

$$w_{T_2} = \frac{f^2 E_0^2}{2c\Omega} N r_0 \left\{\frac{\cos^{2n+1}\psi_T - \cos^{2n+1}\psi_{rim}}{2n+1} + \tan\varphi_T I_3\right\},$$

(12)

Source of losses	Notation	Decrease in A, %
Scattering on reflector rim	w_{rim}/w_p	2.9
Scattering on feed	w_{sh}/w_p	2.8
Scattering of plane wave on braces	w_{T_1}/w_p	3.8
Scattering of spherical wave on braces	w_{T_2}/w_p	10.2
Total losses	—	19.7

where

$$I_3 = \int_{\psi_{rim}}^{\psi_T} \cos^{2n-1}\psi \, d\psi$$

is graphically integrated at specified n, ψ_{rim}, and ψ_r. The conversion from the total power radiated by the aperture to the power measured in the main direction (main lobe), which interests us here, may be performed using the formulas given by Kuznetsov in [6]. In his calculations, the case where many wavelengths are accommodated over the aperture cross section shows only 4% of the power emitted outside the main beam. This shows why we can use the above ratios of scattered power to w_p (this last term referring to the power radiated in the main direction in the absence of scattering on the rim or on bracing bars) to estimate the loss in directive gain in the main direction.

Calculations based on formulas (4), (6), (7), and (10) lead to the results in the table above, in the case of the parabolic RT-22 radio telescope antenna.

The accuracy of this estimate will probably not exceed 10%. The relatively large contribution made by scattering of a spherical wave on bracing bars, attaining 10% despite the field loss toward the rim, is conspicuous. Practically half the scattering losses are accounted for by this source.

More effective use of a reflector antenna can be attained, of course, by a more efficient aperture feed using multireflector arrays [7, 8]. The noise temperature of the antenna is reduced in the process. But any increase in the field on the aperture periphery leads, as we see, to increased scattering of spherical waves on the bars.

A slight reduction in scattering on bracing bars may be achieved by bringing the vertex of the pyramid formed by the bracing bars as far out as possible beyond the reflector focus. But it is still difficult to effect any substantial reduction in scattering by this approach. A radical approach would be to increase the efficiency of a reflector antenna by placing the bar supports on the reflector rim. A successful handling of the problem would however require painstaking calculations of the additional elastic deformations of the reflector brought about by the loading of the rim by the bracing bars.

Only under conditions where the load of the bracing bars on the reflector rim will cause no appreciable increase in reflector deformations and a concomitant decrease in effective area would this approach be permissible. Otherwise the gain achieved by reduced scattering would be more than offset by the decrease in effective area through the resulting increased deformations.

It is precisely owing to the absence of such calculations in the design and erection of the RT-22 radio telescope that we were obliged to place the load of the bracing bars on force-bearing elements of the supporting structure which are practically immune to deformation. At the present time, the necessary calculations have been performed; the results will be given in the subsequent sections, with the design solution based on them.

Loading of Bracing Bars on Reflector Rim and Deformation of Rim

As mentioned earlier, scattering on the bracing bars can be reduced substantially by loading the bars on the reflector rim. But this design can be realized only when the reflector surface will not be subject to excessive deformation. From this standpoint, the load-bearing bar structures must be as light in weight as possible.

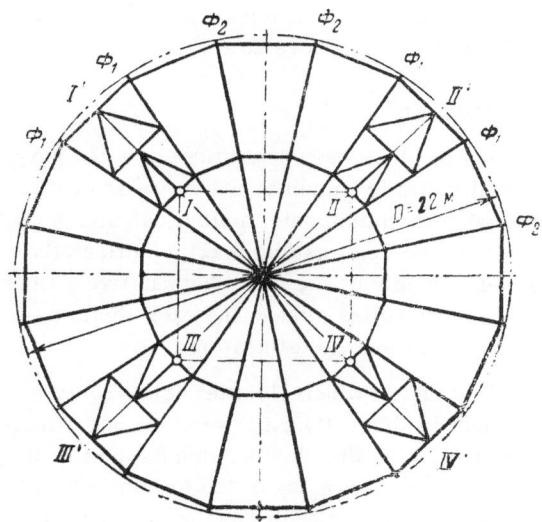

Fig. 2. Structure of force-bearing frame of RT-22 radio telescope reflector (reflector supported at four points). I, II, III, IV are support points of reflector; I', II', III', IV' are points where spars of the pyramid (formed by bracing bars) are joined to reflector rim.

And they must be minimized in size, of course. But reduction in the weight and size of the bar structures entails reduction in rigidity, and this in turn implies increased elastic displacements of the feed, which must remain at the phase center of the reflector antenna when the reflector is in its operating positions. The problem of reducing scattering on bracing bars must be solved taking account of allowable deformations of the reflector and allowable elastic deformations of the bar structures leading to displacements of the feed.

Usually, the structures holding bracing bars in place form a pyramid and are situated symmetrically about the points of support of the antenna on the supporting and rotating pedestal. This arrangement is dictated by an effort to achieve symmetrical loading of the antenna by the bars, and at the same time eliminate nonuniformity of elastic deformations of the antenna reflecting surface. The number of bars forming the bracing structure must be some multiple of the number of points on which the antenna rests.* The requirement that the weight of the bracing structures be minimized (and that the weight of the feed and the equipment placed near the focus be minimized) is inessential only in those cases where the bars are not supported on the force-bearing elements of the reflector itself. This requirement can be lifted, for example, in the design of the RT-22 radio telescope: four bars of the bracing structure bear on the four antenna suspension structures. We must now take these requirements into account in view of the need to place the joints of the bars and reflector rim on the rim. A recent study of the rigidity of the RT-22 radio telescope reflector antenna led to an interesting conclusion which opened up the way for resolving this difficulty successfully. It was found that when the antenna was supported at four points, the joints of the bracing bars on the reflector rim could be placed on the rim with a certain limitation on the weight of the feed system — not only with no deterioration in the geometry of the reflector surface, but even with some improvement through balancing out of deformations. The point is that the force-bearing structure pattern in the RT-22 antenna is not one of perfect radial symmetry (Fig. 2). Perfect radial symmetry in the frame of an antenna having 16 force-bearing radial girders would require suspension on eight points of support situated in the gap between every other girder [9]. In the four-point suspension pattern the radial girders Φ_1 situated close to the points of support exhibit less deflection than the girders Φ_2 further removed from points of support. An increase in load on the closer force-bearing girders within reasonable proportions, increasing their deflection to that of the more remote girders, would therefore smooth out and make the distribution of deformations uniform.

Static calculations of the antenna frame in the RT-22 radio telescope enabled us to find the allowable additional load due to the bracing structures and feed on the radial girders placed closer to the points of support on which the antenna rests. This loading constituted about 20% of the dead weight of the radial force-bearing girder, or about 6.5% of total weight of the radial elements. A greater than 20% increase in loading would have undesirable effects.

The weight of the bracing structures exerts not only vertically acting forces† but also horizontally acting forces (thrust forces). That is why it is necessary to mount these sections of the beam of the outer annu-

---

*In antennas with radially symmetric support, e.g., the antenna of the Australian 65-meter diameter radio telescope, the number of bars may be reduced to a minimum, viz., three. The number of force-bearing radial elements of the reflector frame is, correspondingly, a multiple of three.
† This assumes the reflector axis is vertically oriented.

lus of the antenna frame outside the antenna aperture, in order to avoid having the antenna aperture plane obstructed by tie rods and the like. The weight of these beams is included in the above 20%.

Deformations of Bracing Structures and Bracing Design

Once the allowable weight of the pyramidal structure (bracing bars) has been determined, we can proceed to determine the size. Here, the allowable deformations resulting in longitudinal and crosswise displacements of the antenna feed are essential. It is important to note that temperature deformations can also have some effect, in addition to deformations by weight loading. Temperature deformations exert a particularly strong influence on antenna parameters when the antenna is heated unevenly by the sun. Comparatively little study of temperature deformations in antennas has been made. The results cited refer to the case where these deformations can be safely neglected (assuming uniform heating or insulation against heating).

Another important problem is wind loading. Calculations show that at wind velocities below 10-12 m per sec, static wind loading will not be greater than 12% of the weight loading. Dynamic wind loading affecting relatively light bracing structures is more significant. Calculations show that the fundamental period of oscillation of the bars is about 0.25 sec. Wind gusts cannot result in resonant swaying of the bars for that reason. But dynamic loading in self-sustained oscillations due to periodic detachment of eddies behind the bars* will still not cause the stresses in the bar material to exceed the elastic limit, as calculations indicate.

The maximum allowable longitudinal (along the reflector axis) displacements of the feed, Δf_{long}, are given, as in [6], by

$$\Delta f_{long} = \frac{\lambda}{8(1 - \cos \psi_0)},$$

(13)

where ψ_0 is the angular aperture of the reflector antenna. When $\psi_0 = 60°$ and the operating wavelength $\lambda = 8$ mm, for example, $\Delta f_{long} = \pm 2$ mm. In those cases where a two-reflector Cassegrain-type system is in use (e.g., cf. [10]) and the bracing structures support not the feed but a secondary reradiating reflector, the allowable longitudinal displacement will obviously be half that in the previous example, viz., $\Delta f_{long} = \pm 1$ mm.

The allowable transverse displacement Δf_{tr} is determined by the allowable rotation of the electrical axis of the reflector antenna, Δ_1. In many cases, the latter must not be permitted to exceed one-fifth of the width of the main antenna lobe $\theta_{0.5}$. When $f/2R_0 \cong 0.43$, we have $\psi_0 = 60°$. Clearly [6], $\theta_1 = 0.9\theta_2$, where θ_2 is the angular displacement of the feed from the main focus of the parabolic antenna. Recalling that $\theta_{0.5} \approx 1.25 \lambda/2R_0$, we have for this case

$$\Delta f_{tr} = 0.12 \lambda.$$

(14)

When $\lambda = 8$ mm, we have $\Delta f_{tr} = \pm 1$ mm. Rigidity calculations of the brace structures† include the two cases: a) symmetrical loading (aperture plane of reflector antenna horizontal), b) skew-symmetrical loading (aperture plane vertical).

In the first case, the bars are deflected owing to the transverse component of the dead weight load symmetrically with respect to the apex of the pyramid. The joint at the pyramid apex O is displaced in the direction of the reflector axis longitudinally, with no rotation (Fig. 3). In the second case, the bars are deformed by their dead weight in such a way that the joint O is rotated through an angle θ_0 in the vertical plane symmetric about the rods forming the pyramid (Fig. 4). In the first case, which is of no particular interest as regards technique, we get

$$\Delta f_{long} = \frac{P_j l}{4EF \cos^2 \alpha},$$

(15)

---

*Periodic formation and detachment of eddies occurs aft of a cylindrical rod swept by an air stream.

†For definiteness, we consider a quadripod structure carrying a bar for mounting the secondary reradiating reflector of the Cassegrain system.

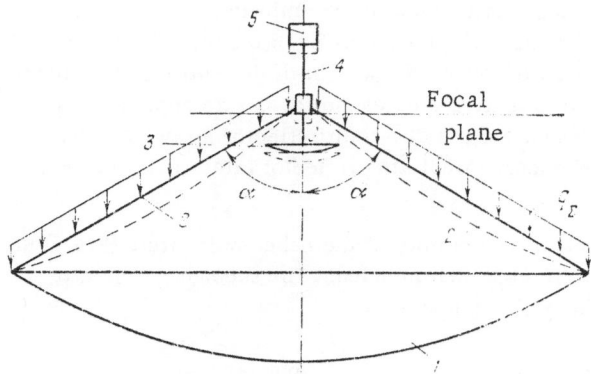

Fig. 3. Mounting plan for a quadripod pyramid (bracing structure) supported on reflector rim (reflector axis oriented vertically). 1) Parabolic antenna; 2) quadripod load-bearing pyramid; 3) small reradiating reflector; 4) rod for small reflector; 5) counterweight. Broken curve indicates deformed position of structure. q_Σ is distributed load, due to dead weight, on rods.

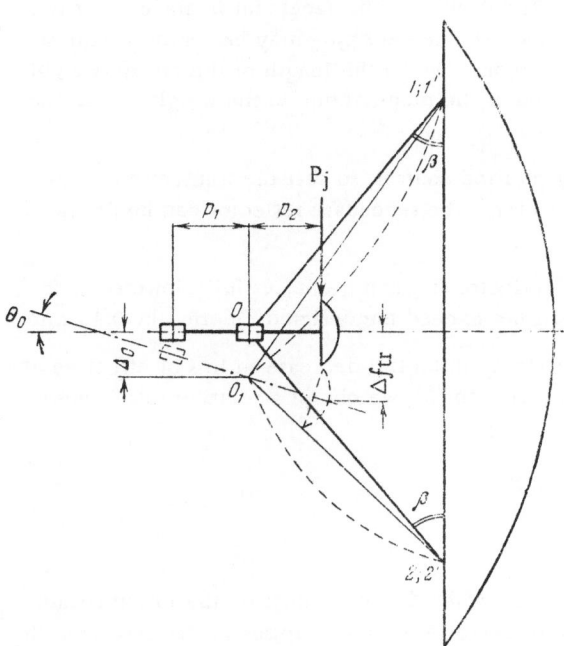

Fig. 4. Diagram of transverse deformations of the load-bearing quadripod pyramid (brace bars) with reflector positioned horizontally. 1', 2, 2') Points where bars forming the pyramid are joined to the reflector rim (broken curve indicates position when deformed).

where P_j [kg] is the load on the joint O [kg has the dimensionality of force], l [cm] is the length of a rod in the pyramid, E [kg/cm^2] is the elastic modulus of the material forming the structure, F [cm^2] is the cross-section area of the rod, 2α is the apex angle of the pyramid (the angle formed by rods situated in the diametral plane). As we see from Eq. (15), Δf_{long} increases rapidly as the angle α increases. This will occur when the bar supports are placed on the reflector rim.

In the second case, the displacement is purely a transverse one and is given by the formula

$$\Delta f_{tr} = \frac{P_j l}{4EF \cos^2\varphi \cos^2\beta} + p_2\theta_0 = \Delta_v + p_2\theta_0, \quad (16)$$

where φ is the angle between adjacent bars; β is the dihedral angle formed by the plane passing through rods 01 and 01' and the reflector aperture (Fig. 4); p_2 is the distance from the secondary reflector to the apex O of the pyramid; θ_0 is the rotation angle of the joint O. The first term in (16) accounts for longitudinal deformation of the bars, the second for displacement of the secondary reflector by the transverse deformation of the bars.

$p_2\theta_0$ is usually almost one order of magnitude greater than Δ_0. But it may be offset entirely in practice by introducing a counterweight. The compensation of Δ_0 cannot be achieved by simple means, nevertheless. Determination of Δ_0 presents no particular difficulties and the procedure is similar to the calculations in the first case. To determine $p_2\theta_0$ we require static calculations of the quadripod frame in space. Calculations based on the RT-22 antenna yield $\theta_0 = 33.5 q_\Sigma/I$ (where q_Σ is the distributed load acting on the bar and I is the moment of inertia of the bar).

Since the cross section F through the bars is a function of the load P_j, while the load itself is a function of the bar's weight, i.e., a function F, the problem of finding the dimensions of the structure is solved by the method of successive approximations. Rough estimates suggest a choice of F = 3.6 cm^2. We then have I $\approx$ 65F = 235 cm^4.* Such a large moment of inertia for a relatively small cross-sectional area of the bar may be obtained only when a thin-walled hollow rod is used.

---

* The conditions for allowable bar flexibility give $l\sqrt{F/I} \leq 160$; hence I $\geq Fl/160$, where l is the length of a bracing bar. In the RT-22 telescope, $l = 1247$ cm.

The bars are therefore made as stainless steel cylindrical shell structures 230 mm in diameter fastened by crosswise diaphragms spaced every 60 cm. The wall thickness 0.5 mm yields F = 3.6 cm^2. The linear weight of the bar q_Σ = 3.9 kg/m; then θ_0 = 0.006. The need to reduce the weight of the entire system bearing on the reflector, plus the problem of how to increase the effective area of the radio-telescope reflector antenna by more efficient use of the feed pattern, tend to favor the use of a multireflector array. As mentioned earlier, then, the bracing structure will carry the secondary reradiating reflector rather than the feed and the high-frequency equipment.

The bracing structure as developed at the radio astronomy laboratory of the Lebedev Institute (FIAN) incorporates aluminum sheet and aluminum alloy. Joints are riveted and threaded. The secondary reflector weighs 80 kg (force units). Below we list the weights of the basic components:

	Wt., kg f
Secondary reflector	80
Reflector-bearing pole	30
Joint at pyramid apex	30
Total weight of counterweight, including fasteners	150
Pyramid (bracing) spars	220
Total	510

When the total load on the pyramid apex point P_j = 400 kg, the longitudinal displacement of this joint (and therefore of the secondary reflector) Δf_{long} = 0.6 mm. The transverse displacement is made up of two components (16), so that Δ_0 = 0.5 mm and $p_2\theta_0$ = 4.6 mm. The displacement $p_2\theta_0$ may be removed almost entirely by using a counterweight to set up a compensating torque. When the length of the counterweight moment arm p_1 = 140 cm, its total weight (taking compensation of the moment due to the weight of the secondary reflector into account) will be 136 kg.

The design of the counterweight mounting provides for position control, so that the transverse displacement of the secondary reflector can be compensated more exactly. The secondary reflector can be displaced by adjustment in the axial direction.

Since the weight of the entire system supported on the reflector rim can be successfully lowered to a very modest value, additional deflections at the bar joints will not exceed the design deflection by 0.15 mm.

The design developed here solves the problem of effecting a substantial decrease in loss of effective antenna area by scattering on bracing structures. In effect, according to (5), we obtain a relative loss of power by scattering on bracing bars.

$$\varkappa_2 = \frac{w_T}{w_p} = 6.2\%$$

for the previous aperture feed pattern (q = 0.32).

Comparison of this value and the total losses on bracing bars when the latter bear on the intermediate force-bearing chord (3.8% + 10.2% = 14%) yields a 2.3-times decrease in scattering losses while preserving the shape of the reflector. The total scattering losses are reduced from 19.7% to 11.9% in the process, i.e., almost halved. The conversion to a multireflector array, which makes it possible in principle to achieve more uniform feed of the aperture, provides even greater advantage.

Literature Cited

1.   B. E. Kinber, Radiotekhn. i Elektron., 7(1):90 (1962).
2.   A. I. Potekhin, Some Topics in Diffraction of Electromagnetic Waves, Soviet Radio Press (1948).
3.   F. E. Borgnis and C. H. Papas, Randwertprobleme der Mikrowellenphysik, Springer Verlag (1955).
4.   P. D. Kalachev and A. E. Salomonovich, Radiotekhn. i Éleketron. 6(3):422 (1961).

5. P. D. Kalachev and A. E. Salomonovich, Tr. Fiz. Inst. Akad. Nauk SSSR, 17:13 (1962).
6. G. Z. Aizenberg, Microwave Antennas, Communications Press (1957).
7. L. D. Bakhrakh, Doctoral Thesis, Moscow (1958).
8. L. D. Bakhrakh and I. V. Vavilova, Radiotekhn. i Élektron., No. 7 (1961).
9. P. D. Kalachev, Izv. Vuzov., Mashinostr., No. 12 (1963).
10. A. E. Salomonovich and N. S. Soboleva, Radiotekhn. i Élektron., 4(5):799 (1959).

MEASUREMENT OF NEAR-FIELD PARAMETERS
OF PENCIL-BEAM ANTENNAS

A. E. Salomonovich, B. V. Braude, and N. A. Esepkina

Introduction

Measurement of the parameters of pencil-beam antennas in the near field is a topic of paramount interest at this time. The ratio of the linear dimensions of the aperture to the wavelength in modern antennas designed for radio astronomy, space radio communications, and other communications has been increasing at such a rapid rate that the use of conventional antenna measurements techniques with external land-based transmitters or receivers has become impossible: the distances over which these transmitters or receivers would have to be spaced in order to perform far-field measurements are commensurate with or far in excess of the distance to the horizon. In the case of large antennas, direct use of the method of artificial heat sources [1], in which standard sources of known brightness temperature must be situated in the far field so that their apparent angular diameters will comprise a considerable fraction of the main lobe in the radiation pattern of the antenna being measured, also runs into some difficulties. When the distances at which the far field begins attain tens of kilometers or more, the linear dimensions of artificial sources become prohibitively large.

Because of these difficulties, increasing favor is now being shown to methods for measuring antenna parameters using naturally occurring sources of radio-frequency emission in outer space, as clearly situated in the far field [2]. But as the directivity of antennas is increased further, and particularly in view of the conversion to shorter wavelengths (centimeter and millimeter wavelength bands), the use of even these techniques runs up against some basic difficulties. The problem is that reliable measurements of antenna beam patterns and determination of their parameters require that the width of the main lobe in the antenna pattern being measured be larger than the "apparent" angular dimensions of the source in space used as the far-field transmitter. Furthermore, this source must establish an emission flux density incident on the antenna high enough for the emission to be recorded reliably against the background of intrinsic noise in the receiver coupled to the antenna output.

As the width of the antenna beam pattern narrows down, the number of sources suitable for antenna measurements dwindles. The brightest sources — the sun and the moon — have apparent angular dimension $\approx 30'$ and are directly unsuited to measurements of antenna parameters if the width of the main lobe in the beam pattern is several angular minutes. The most intense discrete sources (Cassiopeia-A, Taurus-A, Virgo-A, and the nebulae in Omega and Orion in the Northern hemisphere) have angular dimensions 4-7' (on the half-brightness level). Only the intense source Cygnus-A has angular dimensions of about 1'. With the exception of Cygnus-A then, discrete sources cannot be employed directly in measurements of antennas where the width of the main lobe is several angular minutes.

There remains one more hindrance to the use of outer space sources in antenna measurements. As we know, the flux density S of most intense space sources of radio-frequency emission drops as the wavelength becomes shorter

$$S \sim \lambda^n,$$

where n > 0. For example, in the case of the intense source Cygnus-A in the wavelength range $\lambda < 20$ cm, the spectral index n is 1.25. Even the most intense discrete sources usually generate insufficient power at the input of modern receivers coupled to antennas for use in antenna measurements in the short wavelength centimeter and, a fortiori, millimeter ranges. Only Jupiter and Venus can be used directly in practical measurements of antenna parameters in the millimeter wavelength range where the antenna beam width is several minutes; Venus can be used only in periods when it approaches the earth.

These difficulties may be overcome by applying the method of adjustment and measuring near-field antenna parameters [3]. Sufficiently precise measurements of the parameters of large antennas can be achieved by combining near-field measurements with the above techniques utilizing space sources and artificial sources. It is important in the solution of this problem to clarify how great the differences are between beam patterns taken in the near field and in the far field when the antenna is focused. As will be clear from the sequel, knowledge of the differences in both major lobe and minor lobes is important. Furthermore, it is important to find out the relations between the gains measured in the far field and in the near field.

The possibility of near-field adjustments of large antennas has been discussed in the literature [3-7]. An experiment using this method for millimeter-wavelength measurements may be found in [8] as well. But in all these papers, except for [4, 6], attention is focused on major-lobe measurements. The present article deals with research on near-field measurements of side lobes and gain of pencil-beam parabolic antennas with beam patterns whose width runs in several minutes. It is precisely such antennas, as stated earlier, which most require measurement of near-field characteristics.

This paper also considers a method for measuring antenna gain in which use can be made of outer space sources whose angular dimensions are larger than the width of the major lobe measured beforehand in the near zone. These methods are illustrated by the results of measurements of the gain of the 22-m RT-22 radio telescope of Lebedev Institute in the millimeter wavelength range.

Comparison of Beam Patterns Measured in Near Field and in Far Field

Consider a paraboloid of rotation. We shall determine its beam patterns in the near field and in the far field, and make a comparison. The beam pattern is to be determined by its field in the aperture, since we are interested here in a small angular region ($\sim$5-10°) about the main direction.*

The beam pattern is determined using the following approximate formula [9]:

$$\overline{E} = \frac{jk}{2\pi} \int_S [[\overline{n}\overline{E}_0]\,\overline{r}]\,\frac{e^{-jkr}}{r}\,dS, \tag{1}$$

where $\overline{E}$ is the field at the point of observation, $\overline{E}_0$ is the field in the antenna aperture, $\overline{n}$ is the normal to the aperture, S is the aperture area. Formula (1) is valid beginning at distances when induction fields can be neglected, i.e., at $1/kr \ll 1$.

For the zone we have, in the amplitude factor

$$\overline{r} \approx \overline{r}_0, \qquad \frac{1}{r} \approx \frac{1}{r_0},$$

and in the phase factor

$$r = r_0 + (\overline{r}_0\overline{\rho}) = r_0 + \rho \sin\theta\,\cos(\varphi - \psi).$$

Here, ρ, φ are the coordinates of the point in the aperture plane; r_0, θ, ψ are the coordinates of the point at which $\overline{E}$ is computed, and $\overline{r} = \overline{r}_0 + \overline{\rho}$.

$$\overline{E} = \frac{jk}{2\pi}\,\frac{e^{-jkr_0}}{r_0}\int_S [[\overline{n}\overline{E}_0]\,\overline{r}_0]\,e^{-jk\rho\sin\theta\,\cos(\varphi-\psi)}\,\rho d\rho\,d\varphi. \tag{2}$$

---

*Actually, several dozen lobes will be contained in a cone of 1° in the case of the antennas considered here.

It is shown in [7] that the antenna has to be focused when it is adjusted for the measurements, i.e., a converging beam must be produced. In this case, the near field is determined as follows:

$$\bar{E}' = \frac{jk}{2\pi} \int_{S} [[\bar{n}\bar{E}_1]\bar{r}] \frac{e^{-jkr}}{r} dS, \tag{3}$$

where $\bar{E}_1 = \alpha\bar{E}_0 e^{-j\psi}$ is the field over the aperture of the focused antenna. The variables α and ψ will be defined below.

Clearly,

$$r = \sqrt{r_0^2 + \rho^2 + 2\rho r_0 \cos (\widehat{r_0\rho})} = \sqrt{r_0^2 + \rho^2 + 2\rho r_0 \sin\theta \cos (\varphi - \psi)}.$$

The antenna may be tuned in the Fresnel zone, i.e., at distances such that only quadratic terms need be retained in the phase factor, viz.:

$$r \approx r_0 \left[1 + \frac{1}{2}\frac{\rho^2}{r_0^2} + \frac{\rho}{r_0} \sin\theta \cos (\varphi - \psi) - \frac{1}{2}\frac{\rho^2}{r_0^2} \sin^2\theta \cos^2 (\varphi - \psi) \right].$$

In our case, $\sin^2\theta \ll 1$ (e.g., $\sin^2\theta \approx 0.001$ when $\theta \approx 2°$) so that $\frac{1}{2}(\rho^2/r_0^2)\sin^2\theta \cos^2(\varphi - \psi) \ll 1$. In the amplitude factor, then

$$r \simeq r_0 \left[1 + \frac{\rho}{r_0} \sin\theta \cos (\varphi - \psi) + \frac{1}{2}\frac{\rho^2}{r_0^2} \right].$$

The term $\frac{1}{2}(\rho^2/r_0^2)$ is independent of the angle θ, i.e., it is the same for all the lobes. For distances r_a at which antenna tuning is permissible [7], this term will be negligibly small in beam patterns several angular minutes wide. In effect, as shown in [7], $r_a \geq D^2/50\lambda$ and $r_0 = D^2/20\lambda$, since, when $D = 2\rho_0 = 10^3\lambda$, we have $\frac{1}{2}(\rho^2/r_0^2) \approx 10^{-4}$.

Under these assumptions we get

$$\begin{rcases} r \approx r_0 \left[1 + \frac{\rho}{r_0} \sin\theta \cos (\varphi - \psi) \right]; \\[2mm] \frac{1}{r} \approx \frac{1}{r_0} \left[1 - \frac{\rho}{r_0} \sin\theta \cos (\varphi - \psi) \right]. \end{rcases} \tag{4}$$

The term $(\rho/r_0) \sin\theta \cos(\varphi - \psi)$ increases with the angle θ, i.e., as the number of the lobes in the beam pattern increases. Clearly, from (4), rays are assumed parallel because of the low value of the term $\frac{1}{2}\rho^2/r_0^2$ in the amplitude factor in (3). We therefore put $\bar{r} \approx \bar{r}_0$ in the vector product (3).

Substituting these expressions into (3), we arrive at the following for the near field:

$$\bar{E}' = \frac{jk}{2\pi} \frac{e^{-jkr_0}}{r_0} \int_{S} [[\bar{n}\bar{E}] \bar{r}_0] e^{-j\left[k\rho \sin\theta \cos (\varphi-\psi)+ \frac{k\rho^2}{r_0} -\psi \right]} \alpha \left[1 - \frac{\rho}{r_0} \sin\theta \cos (\varphi - \psi) \right] \rho d\rho d\varphi. \tag{5}$$

Let us now take up variables α and ψ. As stated earlier, the antenna has to be focused for tuning in the near field, i.e., a converging spherical wavefront centered at the point O must be obtained from the antenna (Fig. 1). In the case of a paraboloid, this can be done by bringing the feed out from the focus.

It can be shown that in the case of a paraboloid of revolution whose generatrix is defined as $\xi = \eta^2/4f$, with a feed brought out from the focus, the following condition obtains (Fig. 2):

$$\rho_1 + \rho_2 = a_0 + r_0' + \frac{\eta^2}{2} \left(\frac{1}{a_0} + \frac{1}{r_0'} - \frac{1}{f} \right) + \frac{\eta^4}{8} \left(\frac{a_0 - f}{f a_0^3} + \frac{r_0' - f}{f r_0'^3} \right) + O(\eta^6). \tag{6}$$

If the feed is placed at the focus $(a = f)$, then

$$\rho_1 + \rho_2 = a_0 + r_0' + \frac{\eta^2}{2r_0'} + \frac{\eta^4}{8} \cdot \frac{r_0' - f}{r_0'^3 f} + O(\eta^6), \tag{7}$$

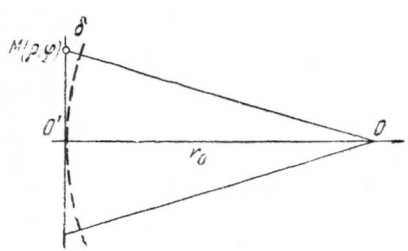

Fig. 1. Converging wavefront.

and as $r_0' \to \infty$, $\rho_1 + \rho_2 \to a_0 + r_0'$, i.e., the antenna is focused at infinitely great distance.

In order to focus the antenna to a finite distance r_0, the feed has to be displaced along the axis to produce the condition

$$\frac{1}{a_0} + \frac{1}{r_0'} - \frac{1}{f} = 0, \text{ or } \quad \frac{1}{a_0} + \frac{1}{r_0'} = \frac{1}{f}. \tag{8}$$

In this case,

$$\rho_1 + \rho_2 = a_0 + r_0' + \frac{\eta^4}{8a_0 r_0' f} + O(\eta^6). \tag{9}$$

For

$$O(\eta^6) \ll \frac{\eta^4}{8a_0 r_0' f} \text{ and } \frac{\eta^4}{8a_0 r_0' f} \ll \frac{\lambda}{2}$$

we obtain*

$$\rho_1 + \rho_2 = a_0 + r_0' = \text{const.}$$

Consequently, a converging spherical wavefront centered at the point O can be produced by bringing the feed out from the focus when a paraboloid of revolution is used.

In antennas of other types, we may indicate alternative methods for focusing at a finite distance r_0. For instance, in a two-reflector Cassegrain array, this can be done by simultaneously displacing the secondary reflector and the feed from the far-field position. It is important to bear in mind that the feed must remain at the focus of the secondary reflector in the course of the displacement (cf. Fig. 3). In this case,

$$FF' = \Delta = \frac{F^2}{r_0}.$$

Formula (8) enables us to find the relationship between the extent the feed is removed from the focus (Δ) and the distance r_0' at which focusing takes place. Substituting $a_0 = f + \Delta$ into (8), we have

$$\Delta = \frac{f^2}{r_0' - f}, \quad \text{since} \quad f \ll r_0', \Delta \simeq \frac{f^2}{r_0'}, r_0' = \frac{f^2}{\Delta}.$$

We then determine the phase distribution and amplitude distribution over the aperture of the antenna so focused.

Clearly, in Fig. 1,

$$r_0^2 + \rho^2 = (r_0 + \delta)^2, \text{ and since } \frac{\delta}{2r_0} \ll 1, \text{ then } \delta \simeq \frac{\rho^2}{2r_0}.$$

That is why the change in phase distribution over the aperture as the feed is moved out is

$$\psi(\rho) = k\delta = \frac{k\rho^2}{2r_0}.$$

When the feed is moved out a distance from the focus, not only the phase distribution, but the amplitude distribution is affected. Indeed, in far-field measurements, i.e., when the feed is at the focus, the amplitude distribution of the field over the aperture is

$$E_0 = \frac{f(\varphi_0)}{\rho_0},$$

---

* This condition gives the permissible distance for adjustment and measurement of the antenna.

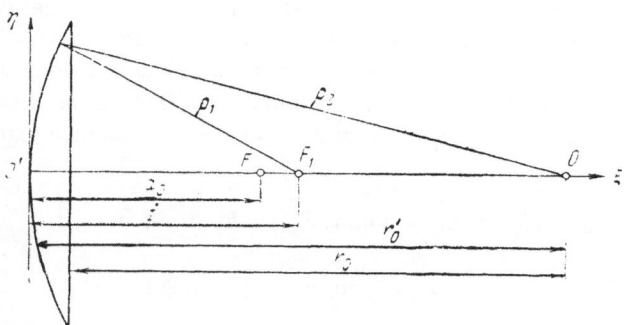

Fig. 2. Illustration for derivation of Eqs. (6)-(9).

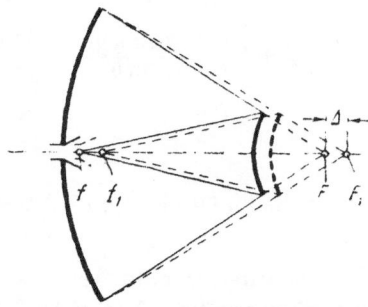

Fig. 3. Focusing of dual-reflector Cassegrain array. F and f are positions of the foci of a parabolic and a hyperbolic reflector with antenna focused at infinity. F_1 and f_1 are the same for focusing at a finite distance by displacing the hyperbolic reflector and feed a distance Δ.

where $f(\varphi_0)$ is the feed radiation pattern, ρ_0 is the distance from focus to reflector, φ_0 is the angle from the focus. When the feed is offset from the focus, the field distribution over the aperture is

$$E_1 = \frac{f(\varphi_0)}{\rho_1} = E_0 \left(1 + \frac{\Delta}{\rho_0} \cos \varphi_0 \right),$$

where

$$\rho_1 = \sqrt{\rho_0^2 + \Delta^2 + 2\rho_0 \Delta \cos \varphi_0} \approx \rho_0 \left(1 + \frac{\Delta}{\rho_0} \cos \varphi_0 \right).$$

For antennas such that $D \geq 10^3 \lambda$ and $r_0 = D^2/20\lambda$, we have $(\Delta/\rho_0)\cos \varphi_0 < 10^{-2}$, and this correction is independent of the angle θ. We may therefore put $E_1 \approx E_0$.

In a focused antenna,

$$\psi = k \frac{\rho^2}{2r_0} \text{ and } \alpha = 1.$$

Substitution of these values of α and ψ into (5) yields

$$\bar{E}' = \frac{jk}{2\pi} \frac{e^{-jkr_0}}{r_0} \int_S [[\bar{n}\bar{E}]\ \bar{r}_0]\ e^{-jk\rho \sin \theta \cos (\varphi - \psi)}\ \rho\ d\rho\ d\varphi\ -$$

$$- \frac{jk}{2\pi} \frac{e^{-jkr_0}}{r_0} \frac{\sin \theta}{r_0} \int_S [[\bar{n}\bar{E}]\ \bar{r}_0]\ e^{-jk\rho \sin \theta \cos (\varphi - \psi)} (\cos \varphi - \psi) \rho^2 d\rho d\varphi = \bar{E} + \Delta \bar{E}. \qquad (10)$$

The beam pattern of a focused antenna, as measured in the near-field zone, constitutes the sum of the far-field pattern and some additional field vanishing in the main direction ($\theta = 0$). This additional term also accounts for the difference in the near-field and far-field patterns. The additional term will be the smaller the further out from the antenna the beam pattern is measured. We find $\Delta \overline{E}$ for a parabolic antenna having a uniform in-phase field distribution over an aperture of circular cross section and diameter $D = 2a$.

In this case,

$$\overline{E}_0 = E_0\,\overline{x};\; E_0 = \text{const}; E_y = E_z = 0; \overline{n} = \overline{z}.$$

$$\overline{E}' = \frac{jk}{2\pi}\frac{e^{-jkr_0}}{r_0}E_0\,(\overline{\theta}_0 \cos\psi + \overline{\psi}_0 \sin\psi \cos\theta) \times$$

$$\times \left[\int\limits_0^a \int\limits_0^{2\pi} e^{-jk\rho\sin\theta\cos(\varphi-\psi)}\,\rho\,d\rho\,d\varphi + \frac{\sin\theta}{r_0}\int\limits_0^a\int\limits_0^{2\pi} e^{-jk\rho\sin\theta\cos(\varphi-\psi)}\cos(\varphi - \psi)\rho^2 d\rho d\varphi \right] =$$

$$= \frac{jk}{2\pi}\frac{e^{-jkr_0}}{r_0}E_0\,(\overline{\theta}_0\cos\psi + \overline{\psi}_0\sin\psi\cos\theta)\,\pi a^2\left\{\frac{2J_1\,(ka\sin\theta)}{ka\sin\theta} + \right.$$

$$\left. + 2\frac{\sin\theta}{r_0}\frac{J_2\,(ka\sin\theta)}{ka\sin\theta}\right\} = A\pi a^2\left\{\frac{2J_1\,(ka\sin\theta)}{ka\sin\theta} + \frac{\lambda}{\pi r_0}J_2(ka\sin\theta)\right\},$$

$$\tag{11}$$

where

$$A = \frac{jk}{2\pi}\frac{e^{-jkr_0}}{r_0}E_0\,(\overline{\theta}_0\cos\psi + \overline{\psi}_0\sin\psi\cos\theta).$$

Clearly, from (11), the near-field pattern differs from the usual far-field pattern by the order of magnitude of λ/r_0. For the pencil-beam antennas discussed here, $\lambda/r_0 \approx 10^{-5}$.[*] Moreover, it is quite evident that as θ increases, the additional term ΔE will first rise and then decline. The maximum $\Delta E = \lambda/2\pi r_0$ occurs in the first side lobe. This is reasonable, since it is obvious that the far field is much closer in the case of side lobes than for the main lobe, and can be measured at relatively short distances from the antenna. Another point is that only the fringes of the reflector surface participate in the formation of side lobes. The foregoing discussion makes it clear that the radiation beam pattern of a focused antenna as measured in the near field is practically identical with the near-field pattern, in the case of pencil-beam antennas, only in the major lobe, but not in the side lobes.[†]

Near-Field Gain Measurements

The antenna gain G is related to the directive gain D and the antenna efficiency η by the formula [10]:

$$G = \eta D. \tag{12}$$

The directive gain may be expressed, in turn, in terms of the radiation directional diagram $F(\theta, \varphi)$:

$$D\,(\theta, \varphi) = 4\pi F^2\,(\theta, \varphi)\left/ \int\limits_{4\pi} F^2\,(\theta, \varphi)\,d\Omega,\right. \tag{13}$$

where integration is carried out over the solid angle $\Omega = 4\pi$. We have

$$G\,(\theta, \varphi) = 4\pi\eta F^2(\theta, \varphi)\left/ \int\limits_{4\pi} F^2\,(\theta, \varphi)\,d\Omega.\right. \tag{14}$$

---

[*]For example, $r_0 \approx 1\text{-}5$ km when $\lambda = 1\text{-}3$ cm [4, 5, 8].

[†]After turning in the manuscript, we became acquainted with an article by J. J. Stangel and W. M. Yarnell (IRE Int. Conv. Rec., 1962, Pt. 1, 3), in which the diagrams of a focused antenna in the near field and the far field are compared, and an expression is derived which is practically the same as ours. The authors of this article appear to be unaware of articles [3-7]. (Note added in proof.)

We restrict our treatment to the gain in the main direction ($\theta = 0$, $\varphi = 0$), and let $F^2(0,0) = f_0^2$,

$$G = 4\pi\eta \left/ \int_{4\pi} \frac{F^2(\theta, \varphi)}{F_0^2} \, d\Omega. \right. \tag{15}$$

We use the notation $F^2(\theta, \varphi)/F_0^2 = F(\theta, \varphi)$, and $F(0,0) = 1$,

$$G = 4\pi\eta \left/ \int_{4\pi} F(\theta, \varphi) \, d\Omega. \right. \tag{16}$$

In most cases of practical importance in the use of pencil-beam antennas, the far-field G must be known in advance.* It is therefore essential to find out to what extent G measured in the near field by any particular technique will differ from G measured in the far field. It is clear from (16) that the answer to this question can be had by comparing the antenna directional diagrams measured in the near field and far field, and that the comparison should be not only of the respective main lobes but of the entire diagram over the solid angle $\Omega = 4\pi$. It is precisely this that dictates the need for detailed treatment of the comparison of side lobes in the previous section. We see from this approach that when the conditions for antenna focusing to a finite distance, as required for near-field measurements, are met, the differences between the near-field and far-field patterns become negligible. The only requirement is that the λ/r_0 ratio be set sufficiently small.

It is vital to dwell on one fact which we have ignored up to this point. As stated earlier, focusing to a finite distance is achieved in a parabolic antenna by moving the feed out axially from the focus. But this operation is known to affect the reflector response to the feed. Generally speaking then, the antenna gain will undergo some change. The question arises: to what degree will the displacement of the feed affect the gain measured in the near zone? It can be shown that moving the feed out from the focus will have a negligible effect on the reflector response to the feed in pencil-beam antennas.

In fact, when the feed is placed at the focus, the reflection coefficient in the transmission line supplying the feed is, on account of the reflector response Γ_1 [11]:

$$\Gamma_1 = \frac{g\lambda}{4\pi f} e^{-j(2kf+b)} = \Gamma_0 e^{-j(2kf+b)}, \tag{17}$$

where $\Gamma_0 = g\lambda/4\pi f$, and g is the feed gain. When the feed is moved out from the focus a distance $\Delta \ll f$, the reflection coefficient Γ_2 is

$$\Gamma_2 = \Gamma_1 e^{-j2k\Delta}. \tag{18}$$

Expressing the gain in terms of the corresponding radiation resistances, and taking the low value of Γ_0 into account, we have

$$G_2 = G_1 \{1 + 2\Gamma_0 \cos(kf+b)(1 - \cos 2k\Delta)\}, \tag{19}$$

where G_1 and G_2 are the gains in the antenna major direction when the antenna is focused at infinity or at a finite distance, respectively.

The maximum relative change in G_1 is, of course,

$$\frac{\Delta G_1}{G_1} = 4\Gamma_0 = g\lambda/\pi f. \tag{20}$$

Since $g \approx 0.6 \cdot 4\pi S/\lambda^2$ (where S is the feed aperture area $\approx \lambda^2$), $\Delta G_1/G_1$ is negligible.

For example, in the case of the RT-22 radio telescope, where $f = 9.5$ m, at wavelength $\lambda = 8$ mm we have $|\Gamma_0| = g\lambda/4\pi f \approx 5 \cdot 10^{-4} = 0.05\%$, so that $|\Delta G''| = 0.002 = 0.2\%$ By focusing the antenna to a finite distance, then, we can measure not only the directional diagram but the antenna gain as well.

---

*The use of pencil-beam antennas for radiometeorological research appears to be an exception.

Basically, the measurement of pencil-beam antenna gain can be reduced, according to (16), to measurement of the directional diagram and to determination of the directive gain. But this is often impossible in practice, requiring measurement of remote side and back lobes over the full solid angle when these lobes have an extremely low level. On the other hand, G_1 [or the effective area $A_1 = (\lambda^2/4\pi)G_1$ in receiver antennas] as determined simply from the major lobe in the case of pencil-beam antennas differs substantially from G found by formula (16).

From the results of the above analysis, we can propose a method using natural sources of radio-frequency emission (in outer space) [2] combined with near-field measurements.

Let

$$\beta_s = \int\limits_{4\pi-\Omega_s} F d\Omega \bigg/ \int\limits_{4\pi} F d\Omega$$

be the scattering coefficient defining the fraction of power radiated into the solid angle outside Ω_s which is the solid angle subtended by the radio emission source in space used as the standard.

When the antenna is pointed on the source, we obtain the power, in terms of the antenna temperature, at the antenna output, as follows:

$$T_{A_1} = [e^{-\gamma}_{\Omega_s}\overline{T}_{b.s} + \overline{T}_{sky}]\,(1-\beta_s)\,\eta + \overline{T}_{bg_{4\pi-\Omega_s}}\beta_s\eta + T_0\,(1-\eta), \qquad (21)$$

where $e^{-\gamma}_{\Omega_s}$ is the absorption in the earth's atmosphere in the direction of the source.

$$\overline{T}_{b.s} = \int\limits_{\Omega_s} T_{b.s}\,(\Omega)\,F\,(\Omega)\,d\Omega \bigg/ \int\limits_{\Omega_s} F\,(\Omega)\,d\Omega\;,$$

[where $T_{b.s}(\Omega)$ is the brightness temperature of the source *],

$$\overline{T}_{sky}\Omega = \int\limits_{\Omega_s} T_s\,(\Omega)\,F\,(\Omega)\,d\Omega \bigg/ \int\limits_{\Omega_s} F\,(\Omega)\,d\Omega,$$

where $T_{sky}(\Omega)$ is the brightness temperature of the sky,

$$\overline{T}_{bg\,\pi-\Omega_s} = \int\limits_{4\pi-\Omega_s} T_{bg}(\Omega)\,F\,(\Omega)\,d\Omega \bigg/ \int\limits_{4\pi-\Omega_s} F(\Omega)\,d\Omega,$$

where $T_{bg}(\Omega)$ is the brightness temperature of the background emission, and T_0 is the temperature of the antenna material.

When the antenna is pointed on the region of the sky adjacent to the source,

$$T_{A_2} = \overline{T}'_{sky\Omega_s}(1-\beta_s)\,\eta + \overline{T}'_{bg_{4\pi-\Omega_s}}\beta_s\eta + T_0\,(1-\eta), \qquad (22)$$

where $\overline{T}'_{sky_{\Omega_s}}$ and $\overline{T}'_{bg_{4\pi}-\Omega_s}$ are the respective brightness temperatures of sky and background as the antenna is displaced. Since, for sources of small angular dimensions, $\overline{T}'_{sky_{\Omega_s}} \approx \overline{T}_{sky_{\Omega_s}}$ and $\overline{T}'_{bg_{4\pi-\Omega_s}} \approx \overline{T}_{bg_{4\pi-\Omega_s}}$,

---

* $\overline{T}_{b.s}$ is the average brightness temperature over the apparent solid angle subtended by the source "weighted" over the antenna directional diagram. In the cases treated here, this quantity may differ substantially from the disk-average brightness temperature of the source.

$$\Delta T_{A_s} = T_{A_1} - T_{A_2} = (1 - \beta_s) \eta e_{\Omega_s}^{-\gamma} \overline{T}_{b.s},$$

and, consequently,

$$(1 - \beta_s) \eta = \frac{\Delta T_{A_s}}{e_{\Omega_s}^{-\gamma} \overline{T}_{b.s}}. \tag{23}$$

Comparing (16) and (23), we get

$$G = \frac{4\pi\eta}{e_{\Omega_s}^{-\gamma}} \frac{\Delta T_{A_s}}{\int\limits_{\Omega_s} T_{b.s} F d\Omega}. \tag{24}$$

We see then that once the distribution of the brightness temperature of the source over the apparent solid angle subtended by that source, as well as the absorption by the earth's atmosphere in the direction of that source, are known, and the antenna directional diagram over the solid angle Ω_s is known and the antenna temperature of the source is found, we can compute the antenna gain. This calls for the use of Eq. (24). It should be emphasized that this procedure is possible in the case of pencil-beam antennas only when data obtained from near-field measurements of the directional diagram $F(\Omega)$ are used.

The source most suitable for such measurements in the centimeter and millimeter wavelength ranges seems to be the moon. The lunar brightness temperature has been measured to satisfactory precision (within 5%) and the brightness temperature over the apparent lunar disk has been studied both theoretically and experimentally as a function of phase and wavelength [12, 13]. The integrated absorption of the atmosphere as a function of height above the horizon, sea-level temperature, and sea-level moisture (clear weather data) has been obtained with sufficient reliability over the entire range of centimeter wavelengths to $\lambda = 0.4$ cm.

The above procedure has been used to find the gain of the RT-22 radio-telescope antenna at $\lambda = 0.8$ mm from data on the near-field directional diagram and from results of lunar radio emission measurements. A matched feed [14] was mounted on the radio telescope, the design of the feed device allowing for removal from the focus to a distance of 25 mm and this corresponds to moving a measuring tower to a distance ≈ 5 km from the antenna. Radiation from a generator installed on such a measuring tower was received at 0.8 cm. Diagrams were taken in the E and H planes by azimuth scanning at variable directions of linear polarization, and the shape of these diagrams was used in adjusting the position of the feed in the focal plane and along the axis of the RT-22 antenna. After the position was optimized, diagrams were recorded within $\pm 16'$ in azimuth in the major lobe and in ten side lobes (five on each side). Figure 4 shows a record in the H plane.

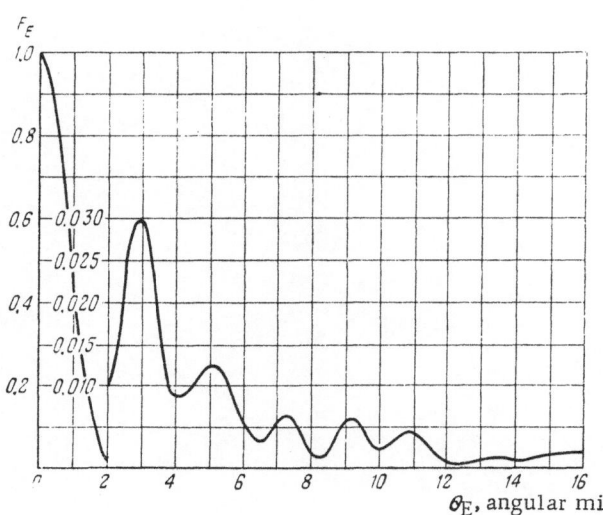

Fig. 4. Diagram in the H plane of the RT-22 radio-telescope antenna ($\lambda = 8$ mm) taken in the near field ($r_0 = 5$ km, $\Delta = 25$ mm).

The gain (or effective area) of the RT-22 antenna was determined as follows. From the known brightness temperature distribution over the lunar disk at 8-mm wavelength at the phase occurring on the day of the observation [12], we made a graphical integration of the integral

$$\int\limits_{\Omega_{moon}} T_{moon}(\Omega) F(\Omega) d\Omega$$

over the solid angle Ω_{moon} subtended by the lunar disk. We used the near-field diagram $F(\Omega)$ with $\pm 16'$. Once the antenna temperature of the moon had been measured, and with the attenuation $e_{\Omega moon}^{-\gamma}$ of the radiation known, we then computed the antenna gain from formula (24). The diagram obtained by averaging the diagrams taken in the E and H planes was used in actual practice. At the major beam width $\varphi_{0.5} = 1'.8$ we found $G = 1.65 \cdot 10^7$. The corresponding effective area is A = 88 ± 10 m^2. When $\varphi_{0.5}$ = 1'.7, we have A = 98.5 ± 11 m^2. Computing the integrals from the diagram over the major lobe and the lunar disk, and assuming η = 1, we were able to find the scattering coefficient outside the major lobe β (this was 0.57) and outside the lunar disk (β_{moon} = 0.24).

Near-field measurements combined with lunar measurements thus enabled us to evaluate the gain and scattering coefficients of the antenna. Effective area measurements in terms of radio emission from Jupiter, whose brightness temperature is assumed 144° on the basis of other available measurements, yield A = 86 ± 5 m^2 at $\varphi_{0.5}$ = 1'.8 and A = 96.5 ± 6 m^2 at $\varphi_{0.5}$ = 1'.7. Measurements by two different methods thus produce results in good agreement, and this confirms the validity of the method proposed.

Literature Cited

1. V. S. Troitskii and N. M. Tseitlin, Izv. Vuzov., Radiofizika, 3:667 (1960); 4:391 (1961).
2. A. D. Kuz'min and A. E. Salomonovich, Radio Astronomical Techniques in Antenna Parameter Measurements, Soviet. Radio (1963).
3. N. A. Esepkina, Dokl. Akad. Nauk SSSR, 113(1) (1957).
4. N. A. Esepkina, Inform. Byul. Leningrad. Politekhn. Inst. Radiofizika, No. 5 (1958).
5. N. A. Esepkina, Pribory i Tekhn. Éksperim., No. 2 (1959).
6. N. A. Esepkina, Doctoral Thesis, Leningrad. Politekhn. Inst. (1958).
7. N. A. Esepkina and V. Yu. Petrun'kin, Inform. Byul. Leningrad. Politekhn. Inst. Radiofizika (1961).
8. A. M. Karachun, A. D. Kuz'min, and A. E. Salomonovich, Radiotekhn. i Élektron., 6(3):430 (1961).
9. V. Yu. Petrun'kin, Tr. Leningrad. Politekhn. Inst. im. M. I. Kalinina. Radiofizika (1955).
10. Centimeter Wavelength Antennas, Soviet. Radio Press (1950).
11. G. Z. Aizenberg, Microwave Antennas, Communications Press (1957).
12. V. D. Krotikov and V. S. Trotskii, Usp. Fiz. Nauk 81(4):589 (1963).
13. A. E. Salomonovich, Astron. Zh., 39(2) (1962).
14. I. V. Vavilova, G. I. Galimov, P. D. Kalachev, A. M. Karchun, A. D. Kuz'min, B. Ya. Losovskii, and A. E. Salomonovich, Vorp. Radioelektroniki, 13(1) (1964).

A SPHERICAL-REFLECTOR RADIO TELESCOPE*

Yu. L. Kokurin and R. L. Sorochenko

The development of radio astronomical research has necessitated the development of large radio telescopes capable of operating in the centimeter and decimeter wavelength ranges. The antenna serving such a radio telescope must have as great a surface area as possible and of rigorously specified form. The precision of the surface determines the minimum operating wavelength and, in consequence, the resolution of the instrument.

Experience in Soviet and foreign radio telescope design shows some serious engineering difficulties in building large antennas with a movable reflecting surface as soon as the reflector area enters the range of several thousand square meters. The problems center on a drastic aggravation of deformation of the reflector accompanying increase in reflector size, with the deformations occurring as the reflector is rotated and traced to the dead weight and the weight of the load-bearing structures. Any attempt to reinforce the latter brings about an increase in detrimental weight loads. Stresses by wind loading also present severe problems.

In this connection, the radio telescope system consisting of a fixed spherical reflector and a small secondary radiator of special configuration placed in the vicinity of the reflector focus seems to hold great promise. The fixed part of the antenna is of small size and can be made in comparatively light weight and with reasonable stiffness. Radio telescopes several hundred meters in diameter can be built in this fashion.

Shape of the Secondary Radiating Surface of a Spherical Radio Telescope

In contrast to a paraboloid, a spherical reflecting surface will not collect the rays incident upon it at a single point. But the spherical aberration taking place can be eliminated by using a secondary reflector to bring the rays to converge after a secondary reflection (Fig. 1). The shape of the secondary reflecting surface may be found by solving the following equations of geometrical optics:

$$\chi = \psi + 2\alpha \qquad (1)$$

condition for reflection from the primary reflecting surface;

$$\tan\frac{\psi}{2} = \frac{d\rho/d\chi}{\rho} \qquad (2)$$

condition for reflection from the secondary reflecting surface;

$$\rho = \frac{cR - 2R\cos\alpha + aR\cos 2\alpha}{\cos\psi + 1} \qquad (3)$$

condition for paths up to point of convergence to be equal, i.e., in-phase condition.

Here, a is a parameter giving the distance between the center and the focusing point selected (in fractions of R); c is a parameter giving the path length traversed by rays from the aperture to the focusing point selected (in fractions of R).

There are four unknowns (χ, ψ, ρ, α) and two parameters (a, c) in the system of Eqs. (1), (2), (3). Differentiation of (3) and substitution of $\psi = \chi - 2\alpha$ from (1) into the denominator yields

---

*Presented to the expanded plenum of the Radio Astronomy Commission in November 1960.

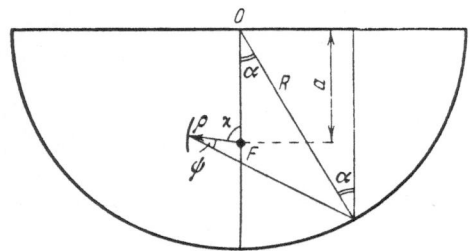

Fig. 1. Path of a ray in a spherical reflector with secondary radiator. O) Center of sphere; F) focusing point of rays; R) radius of sphere; α) angle of incidence of ray on sphere; ψ) angle between ray incident on secondary radiating surface and ray reflected from it; χ) radius vector and polar angle of secondary radiating surface.

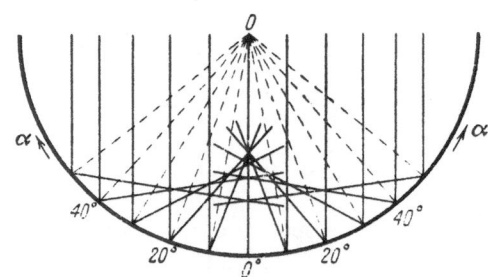

Fig. 2. Convergence of rays reflected from the spherical surface at various angles of incidence α.

$$\frac{d\rho}{d\chi} = \frac{[\cos(\chi - 2\alpha) + 1]\left[2R\sin\alpha\,\frac{d\alpha}{d\chi} - 2aR\sin 2\alpha\,\frac{d\alpha}{d\chi}\right]}{[\cos(\chi - 2\alpha) + 1]^2} +$$
$$+ \frac{[cR - 2R\cos\alpha + aR\cos 2\alpha][\sin(\chi - 2\alpha)]\left[1 - 2\frac{d\alpha}{d\chi}\right]}{[\cos(\chi - 2\alpha) + 1]^2}. \tag{4}$$

From Eq. (1) we have

$$\frac{d\alpha}{d\chi} = \frac{1}{2 + \frac{d\psi}{d\alpha}}. \tag{5}$$

Then, transforming (4) into a function of ψ and α, we get

$$\frac{d\rho}{d\chi} = \frac{[(\cos\psi + 1)(2R\sin\alpha - 2aR\sin 2\alpha)] + \left[(cR - 2R\cos\alpha + aR\cos 2\alpha)\sin\psi\,\frac{d\psi}{d\alpha}\right]}{\left(2 + \frac{d\psi}{d\alpha}\right)(\cos\psi + 1)^2}. \tag{6}$$

Substitution of Eqs. (6) and (3) into (2) yields

$$\tan\frac{\psi}{2} = \frac{[(\cos\psi + 1)(2R\sin\alpha - 2aR\sin 2\alpha)] + \left[(cR - 2R\cos\alpha + aR\cos 2\alpha)\sin\psi\,\frac{d\psi}{d\alpha}\right]}{(\cos\psi + 1)\left(2 + \frac{d\psi}{d\alpha}\right)(cR - 2R\cos\alpha + aR\cos 2\alpha)}. \tag{7}$$

After transforming,

$$\tan\frac{\psi}{2} = \frac{\sin\alpha - a\sin 2\alpha}{c - 2\cos\alpha + a\cos 2\alpha}. \tag{8}$$

Substitution of (8) into Eqs. (1) and (3) yields the following parametric equations specifying the shape of the secondary radiating surface:

$$\chi = 2\alpha + 2\arctan\left(\frac{\sin\alpha - a\sin 2\alpha}{c - 2\cos\alpha + a\cos 2\alpha}\right); \tag{9}$$

$$\rho = \frac{R}{2}\left[c - 2\cos\alpha + a\cos 2\alpha + \frac{(\sin\alpha - a\sin 2\alpha)^2}{c - 2\cos\alpha + a\cos 2\alpha}\right], \tag{10}$$

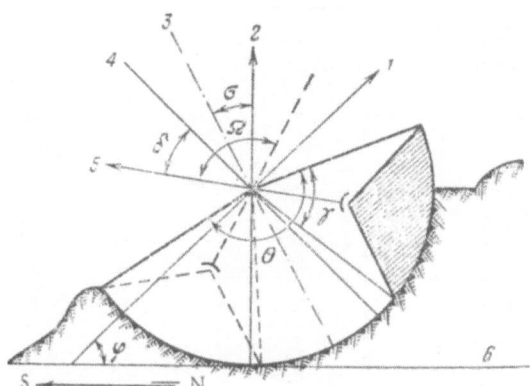

Fig. 3. Coverage of a spherical radio telescope. 1) Earth's axis; 2) zenith; 3) reflector axis; 4) plane of equator; 5) direction of reception; 6) plane of horizon.

where α is the angle of incidence of the ray on the sphere.

Formulas (9) and (10) can be used to compute the secondary reflector parameters, and by varying parameters a and c we can optimize its dimensions and position.

The ray paths seen in Fig. 2 clearly show that: 1) best convergence is obtained at a distance of roughly $R/2$ from the center of the sphere, which is a suitable place for positioning the secondary reflector; 2) the rays incident on the sphere at angles α greater than 35° to 40° begin to diverge markedly and a secondary radiator of very large size is required to intercept them.

Consideration of several variants showed that the secondary reflector can be minimized in size (and this is extremely important when building a large instrument) at $a = 0.5\text{-}0.55$; $c = 1.61\text{-}1.55$. In this case, the diameter of the secondary reflector will be 5-7% of the diameter of the sphere.

Viewing Angle

In a spherical radio telescope, only a portion of the reflecting surface is utilized at one time, so that the direction of reception can be changed by rotating the secondary reflector about the center of the sphere. Observations may be carried out in a cone symmetric about the axis of the spherical surface and of vertex angle Ω. The value of Ω is limited by the central angle θ of the spherical surface (Fig. 3) and by the assigned utilization factor k of the diameter of the sphere. These quantities are interrelated by the formulas

$$\Omega = \theta - \gamma, \; k = \sin \frac{\gamma}{2},$$ (11)

where γ is the central angle of the sphere illuminated by the secondary reflector.

Maximum coverage of the sky with the aid of a spherical radio telescope requires that the earth's axis (extended beyond the poles) be outside the cone of reception, for otherwise the range of possible observations would be restricted in δ, and consequently the number of sources falling in the viewing zone would be diminished. In view of the paramount importance of observations in the plane of the ecliptic, the advantage in locating the instrument in lower-latitude regions is evident. When the instrument is located in moderate latitudes, the axis of the spherical reflector must be tilted with respect to the equator.

The coverage of the spherical radio telescope can be expressed conveniently in terms of the maximum time for tracking an infinitely remote space source at different declinations δ. The formula expressing this has the form

$$t_{\mathrm{h}} = \frac{1}{7.5} \arccos \left[1 - \frac{\cos(\delta - \varphi + \sigma) - \cos\left(\dfrac{\theta - \gamma}{2}\right)}{\cos(\varphi - \sigma)} \right].$$ (12)

Here, δ is the declination of the source; φ is the latitude of the observation site and consequently the angle of inclination of the earth's axis to the horizon; σ is the angle between the axis of the spherical reflector and the zenith direction.

The values of t computed for $\varphi = 45°$, $\sigma = 27.5°$, and $\theta = 165°$ are tabulated below, as found from formula (12).

In the range of values tabulated, the spherical radio telescope possesses an axially symmetrical directional diagram of constant shape whose angular dimensions are determined by the diameter of the sphere and the

Declination	Maximum tracking time, h			
δ°	$k = 0.42$ $\gamma = 50°$	$k = 0.5$ $\gamma = 60°$	$k = 0.564$ $\gamma = 70°$	$k = 0.6$ $\gamma = 74°$
—40	0*	—	—	—
—35	3	0*	—	—
—30	4.1	2,8	0*	0* at δ = —28°
—25	5	4	2.7	2.1
—10	6.7	6	5.2	4.85
0	7.4	6.5	5.8	5.7
+10	7.8	7.05	6.4	6.1
+17.5	7.85	7.2	6.5	6.25
+30	7.1	6.95	6.3	6
+45	6.7	6	5.2	4.85
+60	5	4	2.7	2.5
+65	—	2.8	0*	0* at δ = 63°
+70	3	0*	—	—
+75	0*	—	—	—

*Boundary point where the tracking time is zero at the predicted surface utilization factor. The radio telescope is operating as a passive instrument in this case.

utilization factor of the sphere. Once outside of the cone of reception, the parameters of the diagram and its shape deteriorate and the antenna gain declines.

Comparison of Spherical Radio Telescope with Other Known Variants of Large Antennas

As we see from the table, with the aid of a spherical radio telescope placed at latitude $\varphi = 45°$, at the parameters selected for the calculations, we can observe approximately 80% of the sources (allowing for the diurnal motion of the sources on the celestial sphere) which can be picked up by a completely movable instrument steerable to any point in the sky. Figures may vary slightly for other latitudes. The time of continuous tracking of sources in the case of a spherical reflector is roughly one half the time for a completely movable instrument, but this is no serious drawback, since the permissible values of t are entirely adequate for radio astronomical purposes in the overwhelming majority of cases. At the same time, the limitations on a spherical radio telescope referred to earlier are completely offset by the distinct advantages over other types of large antennas currently known. The following are the salient advantages.

1. Relative simplicity in design (as compared to a movable paraboloid). We stress that the siting of the instrument is of exceptional importance in building a spherical surface of large size. Siting on a mountain is advantageous here, if a natural depression can be found for optimizing the tilt of the axis.

2. Simplicity in operation. The direction of reception is changed and sources are tracked by displacing the secondary reflector in a spherical radio telescope. The reflector suspension system can be designed on either equatorial or altazimuth principles, and the instrument is operated with a very simple motion and relatively light weight of moving parts. For comparison, we could point to a variable-contour antenna [2], where several hundred independently programmed simultaneous motions are required to track a source.

From a radio engineering standpoint, the single component feed of the spherical antenna makes the instrument one of maximum flexibility, particularly when working over a broad frequency range, a rather complicated feat for antennas in which a large reception area is achieved by using several distinct components coupled by high-frequency transmission gear.

3. The spherical radio telescope is more stable to noise than other antenna types, since its feed is completely shielded from direct noise by its secondary reflector, and since the entire antenna is sunk in the ground.

The design features of the instrument (incomplete illumination of the reflecting surface, presence of a secondary radiator) leave room for hope that the spherical radio telescope will have a lower antenna noise temperature. These points are particularly crucial in view of the further improvements in the sensitivity of receiving equipment as a result of the use of maser amplifiers and parametric amplifiers in radio astronomy.

Literature Cited

1. A. K. Head, Nature, 179(4652) (1957).
2. S. E. Khaikin, N. L. Kaidanovskii, N. A. Esepkina, and O. N. Shivris, Izv. Glavnaya Astrofiz. Observ. (Pulkovo), 21:167, 3 (1960).

A RADIO SPECTROMETER FOR POLARIZATION
MEASUREMENTS AT METER WAVELENGTHS

V. A. Udal'tsov

Introduction

The hypothesis of a synchrotron origin for the radiation of the galaxy and most discrete sources has received wide acceptance in recent years. It has successfully been confirmed by a variety of experimental facts. In particular, it provides an explanation for the discovery of linearly polarized radiation from the galaxy and discrete sources. The linear polarization of radiation, which has now been observed for many extraterrestrial sources, results from Bremsstrahlung of relativistic electrons in weak regular magnetic fields. Today there are no other explanations for the presence of linearly polarized radiation. Thus, information on the polarization of radio emission contains evidence on the magnetic fields of the radiating regions. Furthermore, in traversing regions with regular magnetic fields, polarized radiation may experience variations induced by the Faraday effect, so that it may also carry information on the magnetic fields of the media through which it has passed.

In view of this situation, polarimetric measurements offer much interest, particularly measurements of galactic radiation. Polarization was first detected in this radiation in 1955 by Razin at wavelengths of 1.45 and 3.3 m [1, 2], and was subsequently confirmed by the Dutch and British observers [3, 4]. The observations of [3] were carried out at 75-cm wavelength with an ~2° antenna beam. In individual regions of the galaxy, quite high polarization has been detected, up to 10-15%. Similar work has been done by other authors [5-8]. In these papers polarization has not been reported. The conflicting results demonstrate the great complexity of polarization measurements of galactic radiation, and, on the other hand, the need to extend these investigations and to apply new efforts toward a systematic survey, with the aim of making a complete study of the whole celestial sphere at different wavelengths.

We have designed a special radiometer for this type of survey. As the antenna of the radiometer, we plan to use the new cross-type DKR-1000 radio telescope of the Lebedev Physics Institute [9]. The radio telescope is a meridional instrument and has only one polarization [10]. Its transmission band is restricted to the wavelength range 2.5-12 m. For many reasons, these telescope specifications are unsuitable for polarization studies of the galaxy. In the first place, the optimum wavelength range for such observations is 30-100 cm. At wavelengths shorter than 30 cm, the radiant intensity becomes lower, even for the unpolarized component. Moreover, the nonequilibrium component due to polarization becomes comparable with the thermal component, and at wavelengths shorter than 20 cm it becomes even smaller than the thermal component (Figs. 1-3).

Provisional computations indicate that at wavelengths longer than 1 m, polarization should diminish because of the Faraday effect in the radiating interstellar medium. However, there are no reliable observational data for the wavelength range where the transition from polarized to unpolarized radiation occurs. Thus, if we accept the results of [2] as obtained at 3.3-m wavelength, there would be grounds for believing that it might prove fruitful to use the high-frequency range of the DKR-1000 for polarization studies.

In the second place, a stationary or semistationary antenna alone, or a system of such antennas with a single invariant polarization, could not be used for polarization measurements at a fixed frequency. With such an antenna system the radiometer could not distinguish a polarized from an unpolarized signal. To measure the polarization parameters it must be possible either to change the angle between the direction of the electric vector of the radiation and the direction of antenna polarization, or to receive signals with an antenna having

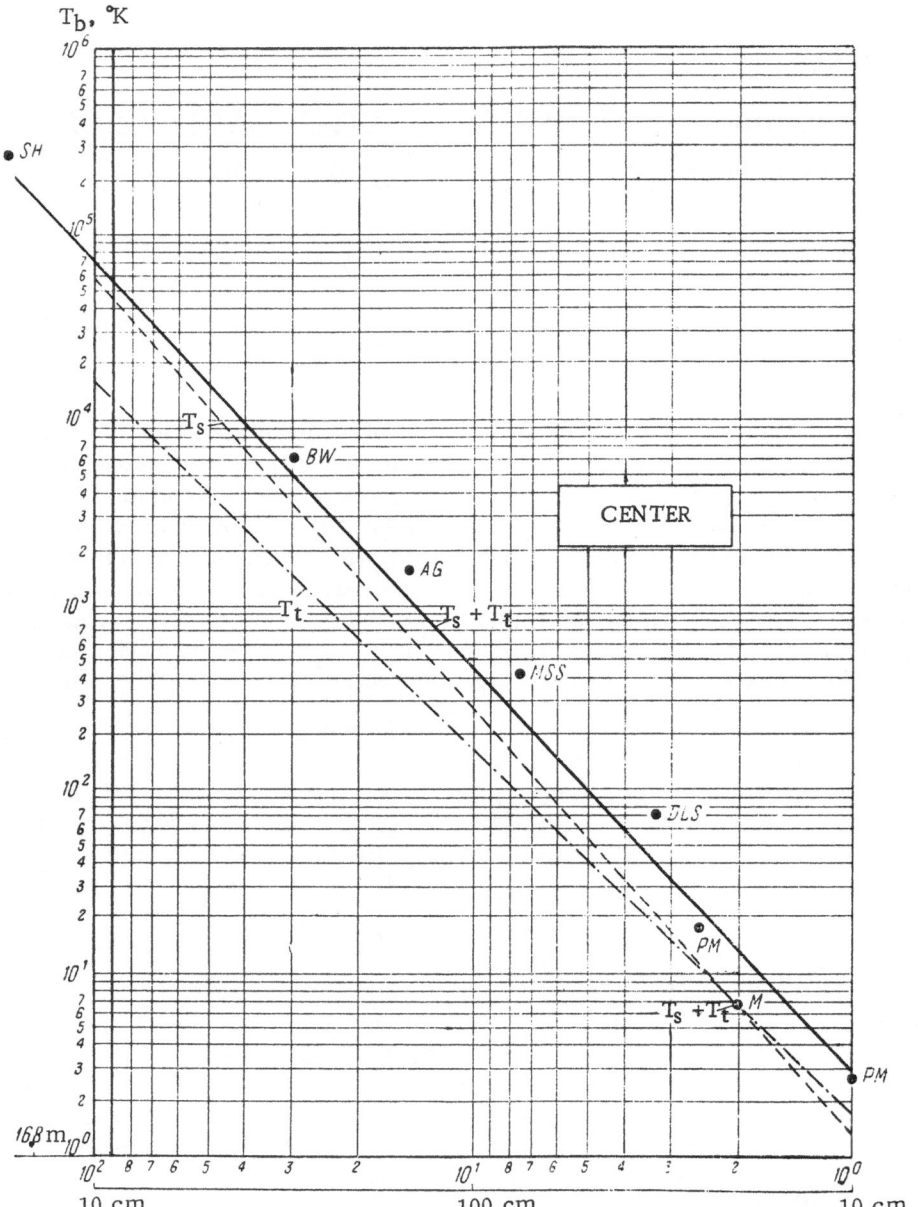

Fig. 1. Radio spectrum for the galactic-center region. T_b) Brightness temperature; T_s) nonequilibrium (synchrotron) component; T_t) thermal component; $T_s + T_t$) total temperature; SH, BW, AG, MSS, DLS, PM, M) experimental points derived by various authors.

two orthogonal polarizations. For measurements at a fixed frequency, the angle can be changed by any of four methods.

1. Simultaneous reception of two orthogonal polarizations [11-13].
2. Successive reception of orthogonal polarizations [14-15].
3. Variation of antenna polarization [2].
4. Variation of the polarization plane of the radiation (time variation of the parallax angle [16].

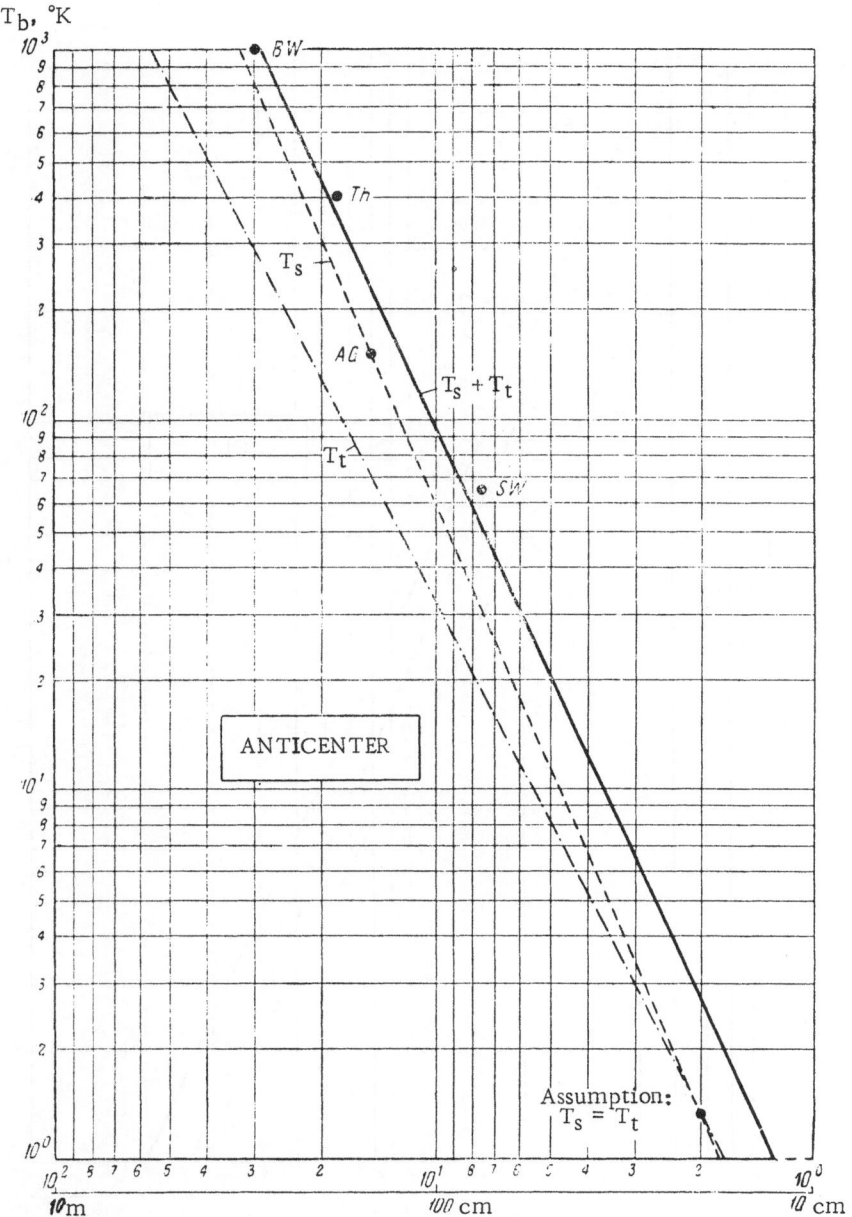

Fig. 2. Radio spectrum for the galactic-anticenter region. Notation the same as for Fig. 1.

It is not possible to apply one of these methods with the DKR-1000 radio telescope; in order to utilize this instrument for polarization measurements, we must find a reliable new technique for varying the angle between the electric vector of the radiation and the antenna polarization plane.

The main difficulties encountered in polarization measurements arise primarily from the fact that a weak polarized signal can be swamped by a strong extraneous signal due to spurious polarization effects. If these effects amount to several percent, as is almost always the case under actual conditions, it will not be possible to measure a polarized component of a fraction of a percent or even one percent. When extraneous polarization effects are present because of the antenna itself and the measuring procedure, the fluctuation sensitivity of the equipment cannot be realized. As a result, one must try to diminish extraneous effects. It would hardly be possible, however, to remove them completely. In order to achieve the full fluctuation sensitivity,then, a method must be found that is capable of distinguishing a true from an extraneous signal.

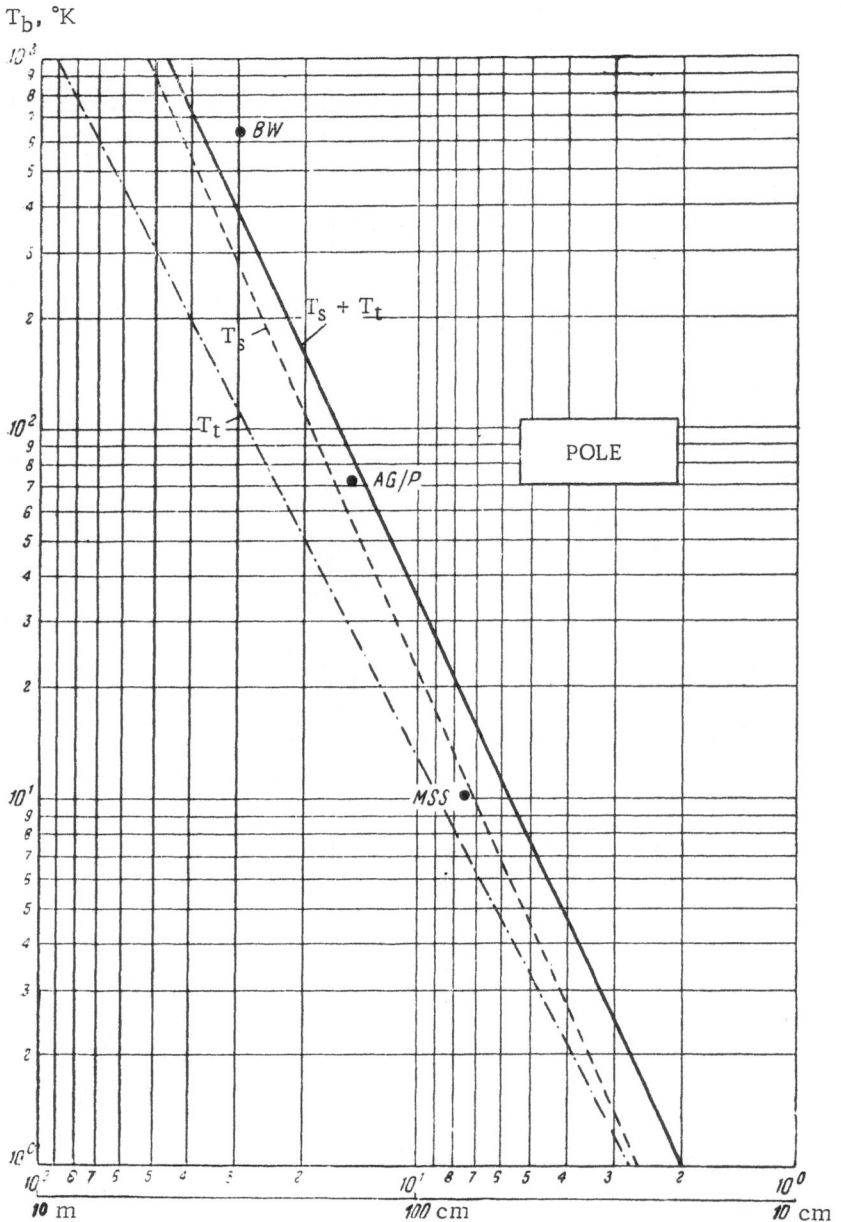

Fig. 3. Radio spectrum for the North Galactic Pole. Notation the same as for Fig. 1.

We should also call attention to the desirability of combining various instrumentation features in a unified receiving apparatus. This would be especially important in developing equipment for survey measurements — the program we are essentially suggesting for polarization studies of galactic radiation. Even if negative results are obtained from the polarization research, the equipment may prove useful for solving other problems, and to a large extent may thereby justify itself from an economic standpoint.

The range of experiments that can be performed in radio astronomy is extremely limited. By radio methods one can basically make direct measurements of four characteristics of extraterrestrial radio emission: 1) the intensity at a fixed frequency; 2) the brightness distribution and angular dimensions of a source; 3) spectral properties; and, 4) polarization properties.

By combining several of these capabilities in a single instrument one can, apart from economic benefits, secure fuller information on sources of radio emission.

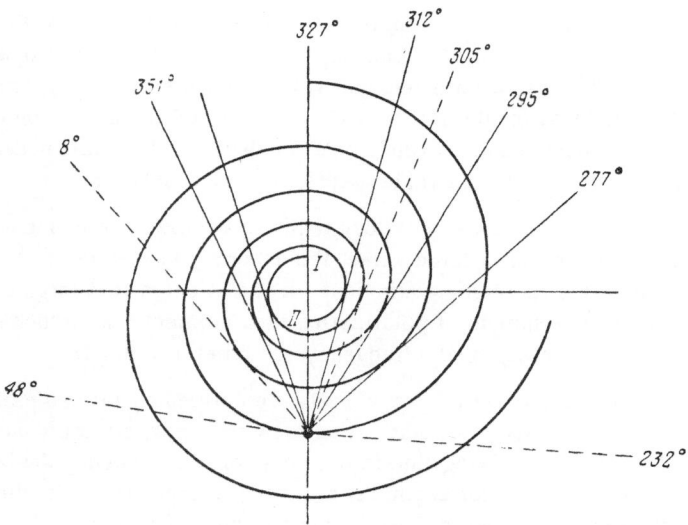

Fig. 4. Pattern of spiral structure in the galaxy, compiled from
B. Y. Mills' data. The directions of longitudinally propagated
radiation (along the spiral arms) are indicated by lines diverging
radially from the sun's position.

Observing Procedure

In order to satisfy all the points mentioned in the preceding section, it is appropriate to use the method
of frequency modulation of a signal. The method depends on the action of a magnetic medium on linearly polarized radiation passing through it (the Faraday effect). The polarization plane of the radiation passing
through a magnetic medium changes by an angle φ which depends on the wavelength λ by

$$\varphi = a_z\lambda^2, \tag{1}$$

where a_z is a parameter for the medium in the z direction, given by

$$a_z = 0.25 \cdot 10^{-16} \int_0^{l_{max}} N_e H_z \, dl_z, \tag{2}$$

with N_e the electron concentration in cm^{-3}; H_z the longitudinal component of the magnetic field, and l_z the
extent of the magnetic medium along the line of site.

Either the interstellar medium or the earth's ionosphere may serve as the magnetic medium. The ionosphere is evidently the more effective. An analysis of the spectrum of the polarized radiation from the Crab
nebula [17] indicates that the rotatory capability of the interstellar medium in the galactic plane (toward the
anticenter) is substantially less than that of the ionosphere ($a_{int.m}/a_{ion} \leq 0.3$).

The regular longitudinal component of the magnetic field is responsible for rotating the plane of polarization in the interstellar medium. Because of the randomized field directions, the cloud structure of the magnetic fields will not affect the orientation of the polarization plane. The maximum regular longitudinal component of the magnetic field should probably occur in the directions of the galactic spiral arms. If the regular
field in the arms has the value 10^{-6} gauss, strong Faraday rotation exceeding the ionospheric value might be observed along these directions. But there are few such directions. If our galaxy has a bispiral structure (Fig. 4),
there would be 16 directions in all (see [18]). Thus, in developing a measuring technique based on Faraday rotation, we should consider primarily the rotation of the polarization plane in the ionosphere.

The frequency dependence of the position angle of the polarization plane for linearly polarized radiation passing through the ionosphere transforms into a frequency dependence of the peak signal for the case where the antenna receives only the polarization for a receiver with resonance frequency tuned within certain limits. A maximum (100%) modulation depth for the intensity of the polarized signal will be observed if the value of the periodic change in the resonance frequency (the frequency deviation) permits periodic measurement of the rotation angle of the polarization plane for an angle greater than or equal to π.

An unpolarized component will be received along with the polarized component if this procedure is followed. If the unpolarized component has a large value, the accuracy of measurement will be small, as the fluctuations in the parameters of the receiving equipment will be affected to a large extent. For this reason it is desirable to use a differential technique of polarization measurement — a method similar to that used for spectral investigations at 21-cm wavelength (the frequency-radiometer method).

If the receiver tuning frequency is varied periodically, a polarized signal can easily be distinguished from an unpolarized one. To raise the accuracy of polarization measurements, attempts should be made to improve the stability of the parameters of the receiving line in the range of the frequency deviation. If the receiver parameters have a frequency dependence (for example, a frequency dependence for the gain), an extraneous signal will be observed at the receiver output due to the unpolarized component.

It is practically impossible to achieve high stability and constancy of the receiver parameters over a wide frequency range for a radiometer with smooth tuning of the resonance frequency. Under these conditions a multichannel reception method is found to be best. In such a system, the channels are tuned to different fixed frequencies. The range of frequency dispersion should correspond to a rotation of the polarization plane by an angle of π or more. For a frequency dispersion over a range corresponding to rotation of the polarization plane by π, one period of signal variation will be observed at the antenna output. The frequency-modulation method mentioned above would be applied to identify the polarized component in each channel. In this system each channel would record the intensity difference between two frequencies. The difference depends both on the frequency dispersion and on the spectral and polarization characteristics of the radiation. With such a system one could derive the dependence of the intensity difference for the received signal on the value of the frequency dispersion. If the difference in the rotation angles of the electric vector at the channel frequencies is a multiple of the channel number, the relation may be expressed in the form

$$y = A(x) + B_0 + B \sin(bx + \varphi_0), \qquad (3)$$

where $A(\lambda)$ is a function describing the spectral dependence of the radiation, x is the serial number of the channel, B_0 is a coefficient given by the position of the polarization plane of the radiation for the greatest of the modulated frequencies (the position relative to the polarization plane of the antenna), b is the periodicity coefficient of the variable component (as given by the dispersion in the modulated frequencies), and φ_0 is the phase of the variable component.

The term $A(x)$ in Eq. (3) specifies the spectral dependence of the radiation received and can furnish a value for the spectral index. The term $B_0 + B \sin(bx + \varphi_0)$ describes the polarized component, whose value is B, with the polarization plane at the maximum modulated frequency given by the coefficients B_0, φ_0. The period dependence corresponding to $x = 2\pi/b$ is not known in advance. It is determined by the rotatory capability of the ionosphere, which depends strongly on time and season. Calculation shows that if the value of n/x, where n is the number of channels, falls in the range 0.5-2, and if the accuracy of measurement of the intensity difference in each channel is δT, then n must be at least 10-12 in order to determine the polarized component to an accuracy of δT. If the range n/x is confined to 0.8-1.2, n may be reduced to 6-10. Figure 5 shows the intensity difference for various positions of the electric vector E_0 as a function of the channel number and the angle ψ; here, ψ is the angle between the electric vector E_0 and the polarization plane of the antenna. The position of the vector E_0 relative to the antenna polarization depends, first, on the orientation of the polarization plane of the radiation and, secondly, on the rotation of this plane in the ionosphere. As the ionosphere parameter a varies, the character of the signal recorded will change.

To determine the complete spectral characteristics of the radiation, we must know the total radiant intensity of the unpolarized component at one of the frequencies in the range investigated. This component is

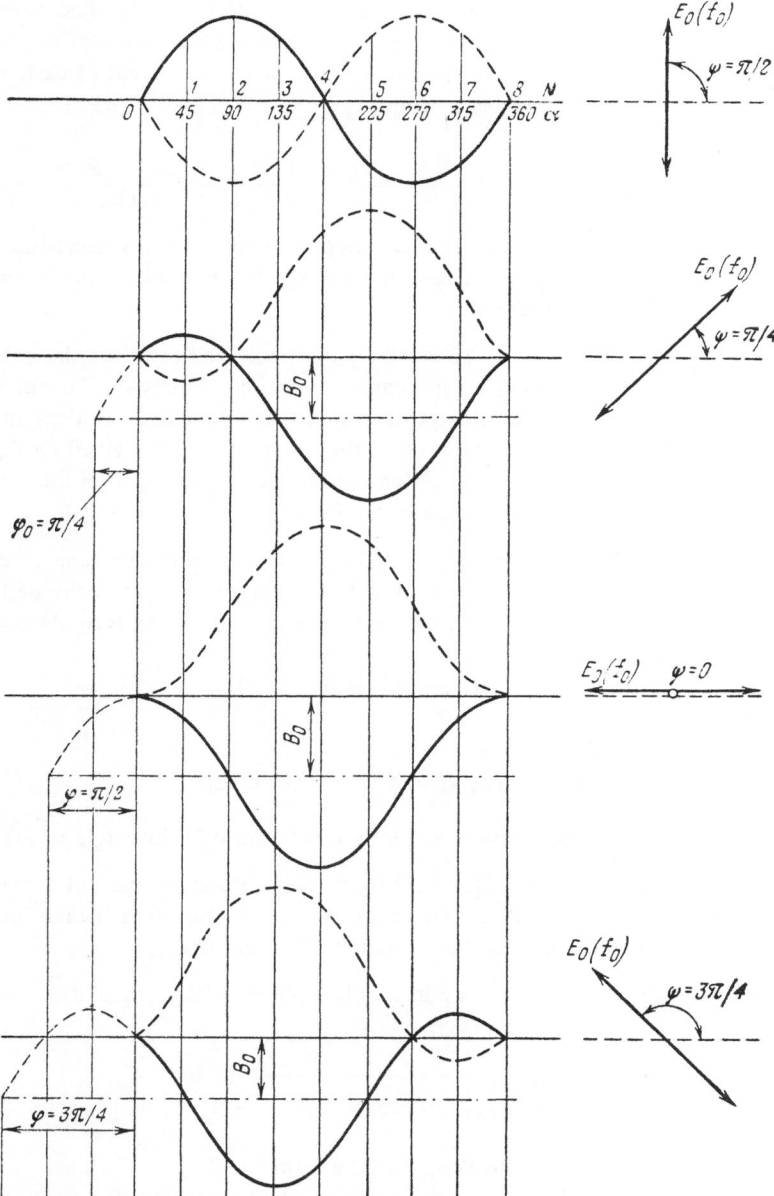

Fig. 5. Intensity difference as a function of frequency dispersion for selected positions of the electric vector of the reference frequency f_0. ψ) Angle between the electric vector E_0 and the polarization plane of the antenna, varying with the rotatory capability of the ionosphere; E_0) electric vector at the reference frequency f_0; N) channel number; α) difference in the rotation angle of the electric vector, observed at the modulated channel frequencies; B_0) term in Eq. (3); φ_0) phase of variable component in Eq. (3). The antenna polarization is shown by broken curves. Whether a solid or broken curve is used depends on the polarity and phasing of the amplitude and phase detectors of the spectrometer channels.

also needed for determining the degree of polarization of the radiation under study. For this purpose, the radio spectrometer should incorporate a channel for recording the total intensity at any of the frequencies received. The best way to record the intensity with the DKR-1000 would probably be to use a correlation method [11, 19, 20]. It would avoid the disadvantages of compensation reception in the absence of modulating systems. More-

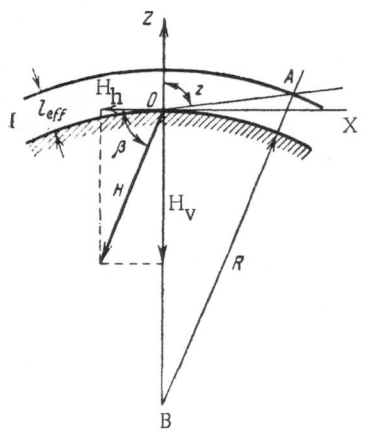

Fig. 6. Derivation of the altitude dependence of the earth's magnetic field. I) Ionosphere; X) plane of horizon; Z) zenith; R) radius of earth; l_{eff}) height of ionosphere in direction of zenith; z) zenith distance.

over, experience has shown it to be the method most resistant to interference; and this circumstance is of particular value, since the DKR-1000 is located near Moscow and the level of various kinds of interference may be high.

Faraday Rotation of the Polarization Plane of Extraterrestrial Radiation in the Ionosphere

Polarization measurements with a spectrometer at meter wavelengths utilize the Faraday effect in the earth's atmosphere, as pointed out above.

To select the parameters and estimate the capabilities of the receiving equipment, we must know the values of the rotation angle of the polarization plane in the ionosphere. In the general case, the Faraday rotation in the ionosphere is described by Eq. (2) and depends on the state of ionization and on the value of the radial component of the earth's magnetic field.

In Fig. 6, if OX and OZ are the directions of the plane of the horizon and the zenith, the radial component of the earth's magnetic field can be represented approximately by the formula

$$H_z = H \cos (\beta - 90° + z) \tag{4}$$

or by

$$H_z = H_V (\cos z + \cot \beta \sin z), \tag{5}$$

where H_V is the vertical component of the earth's magnetic field, and β is the angle of magnetic dip.

The approximation in the expression for H_Z arises from the fact that it does not include the dependence of H on the latitude variation of different ionospheric levels for $z \neq 0$. Nor does it take into account the height dependence of H, or the variations in H induced by currents in the ionosphere.

The dependence of the path length of the ray in the ionosphere on the zenith distance can be determined from the triangle OAB. Here,

$$OA = R \left(\sqrt{\cos^2 z + \frac{2l_{eff}}{R}} - \cos z \right). \tag{6}$$

If we take $l_{eff} = 0.08R$ (R = 12.5), we can write Eq. (6) in the form

$$OA = l_z = l_{eff} 12.5 \left(\sqrt{\cos^2 z + 0.16} - \cos z \right). \tag{7}$$

If we assume that H_Z is independent of height, then by Eqs. (2), (5), and (7) we could write

$$a_z = 0.25 \cdot 10^{-16} H_V B(z) n, \tag{8}$$

where n is the total number of electrons in a column extending the entire thickness of the ionosphere in the direction of the zenith:

$$n = \int_0^{l_{eff}} N_e \, dl; \tag{9}$$

and B(z) is a function of the altitude, given by

$$B(z) = (\cos z \pm \cot \beta \sin z) 12,5 \left(\sqrt{\cos^2 z + 0.16} - \cos z \right). \tag{10}$$

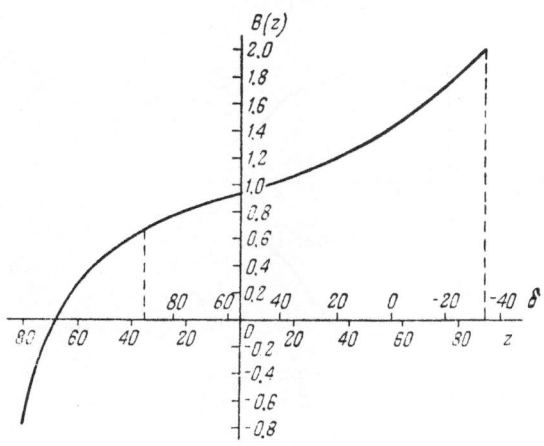

Fig. 7. Graph for the function B(z).

For the position of the Oksk Radio Astronomy Station of the Lebedev Physics Institute, $\varphi = 55°$ and $\lambda = 2^h 30^m$, the approximate value of the magnetic field according to the data of [21] is 0.48 gauss, the magnetic dip $\beta = 60°$, and the magnetic declination $\alpha = 7°$ (epoch 1950.0). For these values of H and β, the horizontal and vertical components of the magnetic field are 0.18 and 0.45 gauss, respectively.

Figure 7 shows the function B(z) for the Oksk Station (epoch 1950.0). The graph indicates that for apparent declinations $\delta = +90°$ to $-35°$ at the latitude of the Station, and for observing conditions at lower culmination (z extending to 35°S and to 90°N), the function B(z) varies by approximately a factor of three. Provisional data for the total number of electrons in a 1-cm² column extending the entire thickness of the ionosphere can be obtained from investigations on the transmissivity of the ionosphere for polarized signals reflected from the moon [22, 23].

Figure 8 illustrates the time variations of the quantity n, as reported by [22, 23]. The following are provisional values of the parameter n for winter and summer, and for night and midday.

	Winter	Summer
Night (toward morning).	$5 \cdot 10^{12}$	$15 \cdot 10^{12}$
Midday.	$20 \cdot 10^{12}$	$60 \cdot 10^{12}$

These data show that the limiting values of the ionospheric parameter n occur at midday in the summer, n_{max}, and on winter nights, n_{min}. Their ratio is 12. If observations at all declinations visible from the station at any time of day and year are contemplated, the receiving equipment should be designed to cover a very large range of variation of the ionospheric parameter a, as given by the ratio

$$\frac{a_{max}}{a_{min}} = \frac{B(z)_{max}}{B(z)_{min}} \frac{n_{max}}{n_{min}} \approx 36 ! \tag{11}$$

For $n_{max} = 60 \cdot 10^{12}$, $B(z) \approx 2$ and $H_v = 0.45$ gauss,

$$a_{max} = 14 \cdot 10^{-4} \ rad/cm^2.$$

For $n_{min} = 5 \cdot 10^{12}$ and $B(z) \approx 0.7$, and the same value of H_v,

$$a_{min} = 0.4 \cdot 10^{-4} \ rad/cm^2.$$

To diminish the range of variation of a, night observations should be rejected. If observations were confined, say, to the period between 8^h and 18^h, the value of a would vary from winter to summer over the range from $15 \cdot 10^{12}$ to $60 \cdot 10^{12}$, and the range of variation of a would be narrowed from a factor of 36 to 12.

Design of the Radio Spectrometer and General Requirements

Figure 9 provides a simplified block diagram for the spectrometer. The diagram shows three receiving channels with frequency modulation and one correlation channel. The number of frequency-modulation channels is governed by the requirements on the accuracy of polarization and spectral measurement. For spectral measurements within the frequency range of the DKR-1000, three to five channels would be needed, distributed with respect to frequency over as wide a range as possible. The circuitry for these channels is similar to that for channels 2 and 8 as shown in Fig. 9. Their frequencies are:

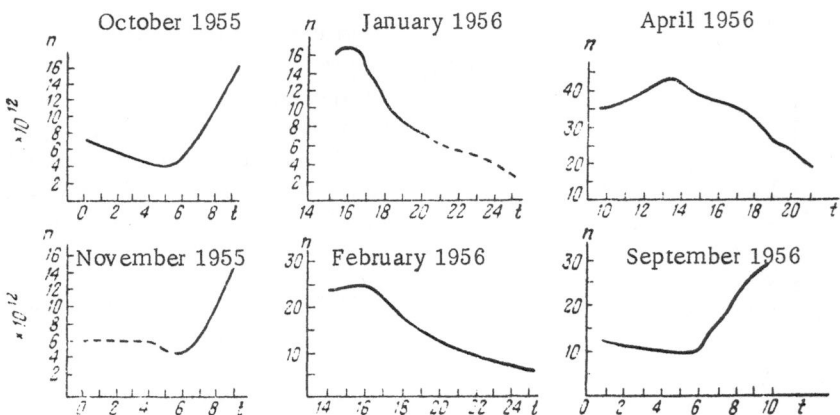

Fig. 8. Time variations of the total number of electrons within the thickness of the ionosphere.

$$f_0, f_9 = 83 \text{ Mc} \quad (\lambda_9 = 3.6 \text{ m}),$$
$$f_0, f_{10} = 60 \text{ Mc} \quad (\lambda_{10} = 5 \text{ m}),$$
$$f_0, f_{11} = 40 \text{ Mc} \quad (\lambda_{11} = 7.5 \text{ m}),$$
$$f_0, f_{12} = 30 \text{ Mc} \quad (\lambda_{12} = 10 \text{ m}).$$

The frequency f_0 is common for all the channels and is equal to 118 Mc (λ_0 = 2.54 m).

Frequency Range of the Polarimeter. Polarization studies of galactic radiation should be carried out, as mentioned above, within a wavelength range of approximately λ = 30-100 cm. For this application, the DKR-1000 would have to be used at the short-wave end of its transmission band, in the region near 120-Mc frequency. The frequency range required for measuring polarization by the method discussed is determined by the requirement that the electric vector rotate by the angle π. If the Faraday rotation of the polarization plane at a fixed frequency is determined by Eq. (1), the difference in the rotation angles of the polarization plane at two frequencies will be

$$\Delta\varphi = a\,(\lambda_{max}^2 - \lambda_{min}^2). \tag{12}$$

If $\lambda_{min} = \lambda_0$ = 2.54 m (f = 118 Mc), then for a weakly active ionosphere ($a = a_{min} = 0.4 \cdot 10^{-4}$ rad per cm^2) a rotation of the electric vector by an angle of π will be observed at a wavelength λ_{max} = 3.75 m (f_{min} = 80 Mc). The maximum frequency dispersion would then be 38 Mc.

Since the ionosphere does not remain in a constant state and varies strongly with time, the receiver permits changes to be made in the range of observed frequencies. For maximum ionosphere activity ($a = a_{max} = 12 \cdot 10^{-4}$ rad/cm^2), the lowest of the modulated frequencies should, by Eq. (12), decrease to 116 Mc (λ_{max} = 2.58 m). The maximum frequency dispersion in this case is $f_0 - f_{min}$ = 2 Mc. For a moderately active ionosphere, the frequency dispersion may be taken as 8 Mc (f_{min} = 110 Mc).

Thus, in each spectrometer channel used for polarization measurements, one of the modulated frequencies f_n', f_n'', or f_n''' may be varied. If the second modulated frequency f_0 is held constant, the frequency difference $f_0 - f_n$ will vary. This difference will be greatest in the last channel and will take the following values:

Maximum	32 Mc
Average	8 Mc
Minimum	2 Mc

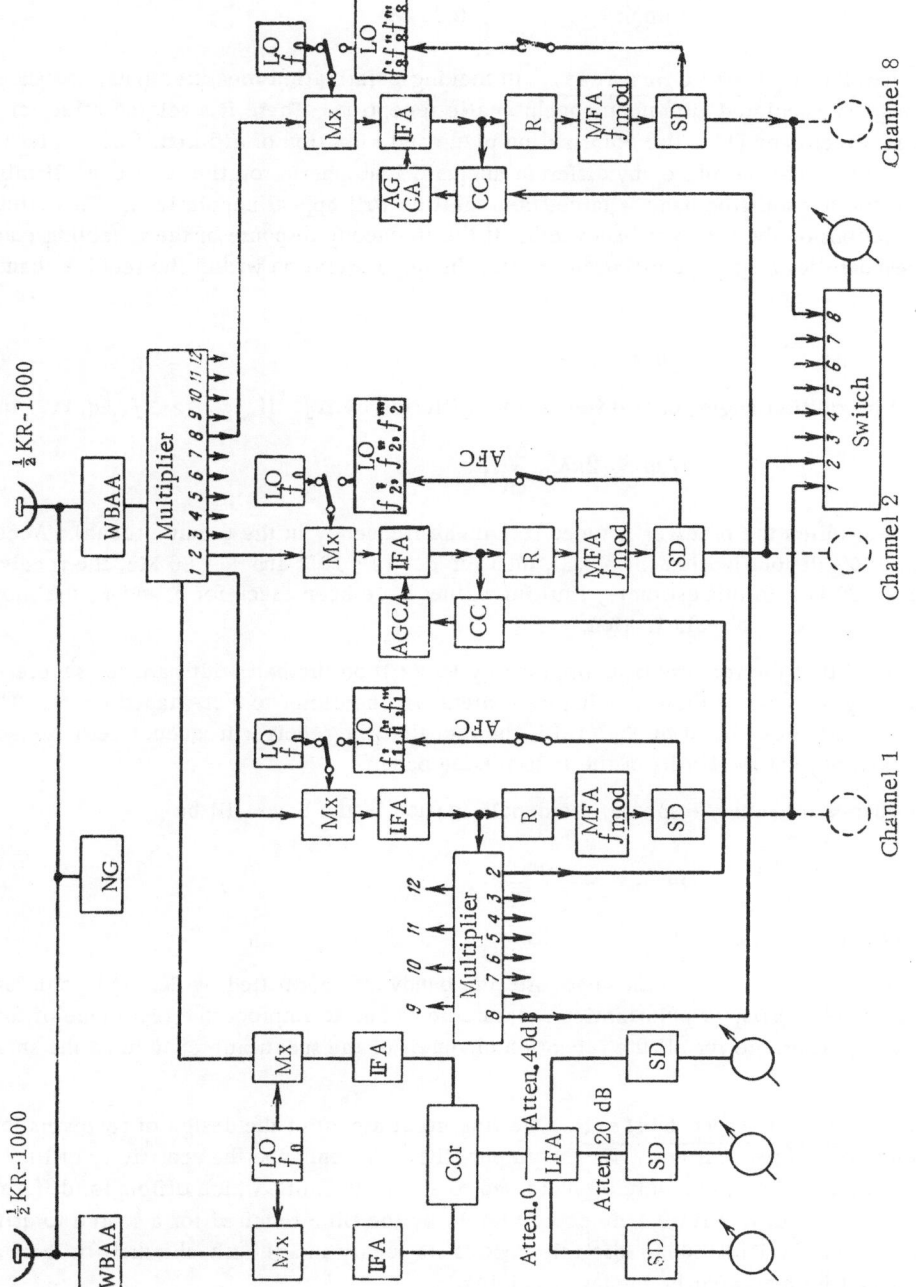

Fig. 9. A simplified block diagram for the radio spectrometer. $\frac{1}{2}$ KR-1000 half of the East—West array of the KR-1000 antenna; NG) standard noise generator; WBAA) wide-band antenna amplifier; Mx) mixer; LO) local oscillator; IFA) intermediate-frequency amplifier; Cor) correlator (multiplying circuit); LFA) low-frequency amplifier; SD) synchronous detector; R) rectifier; MFA) modulation-frequency amplifier; AFC) automatic frequency-response control; CC) comparison circuit; AGCA) automatic gain control amplifier.

To simplify the reduction of the observational results, the maximum difference of 32 Mc has been taken somewhat smaller than the computed value of 38 Mc. For an 8-channel polarimeter, the frequency dispersion in the first channel will have the following values for the ranges given above:

Maximum	$32/8 \doteq 4$ Mc
Average	1 Mc
Minimum	0.25 Mc

Bandwidth of the Reception Channels. In making polarization measurements, one should allow for the extraneous effects associated with nonmonochromatic reception. There is a related effect of signal depolarization in the receiver band [1]. The polarization plane and radiation of different frequencies within the receiver transmission band will be rotated by different angles if ionospheric rotation is strong. If this differential rotation within the transmission band is large, the radiation will appear unpolarized. This circumstance imposes major limitations on the receiver bandwidth. If the frequency response of the receiving equipment corresponds to the response for a single oscillation profile, the depolarization within the receiver bandwidth Δf is given by [1] as

$$m = 1 - e^{-\Delta\Phi}, \tag{13}$$

where $\Delta\varphi$ is the difference in rotation angle for two frequencies differing by Δf. If $f_{av} \gg \Delta f$, Eq. (12) implies

$$\Delta\varphi = 2a\lambda_{av}^2 \frac{\Delta f}{f_{av}}, \tag{14}$$

where f_{av} is the frequency to which the receiver is tuned (the mean frequency in the receiver band). According to Eq. (14), if we admit 20% depolarization (m = 0.2), then for $a = 14 \cdot 10^{-4}$ and $f = 86$ Mc, the receiver bandwidth should not exceed 60 kc. In this estimate, limiting values have been taken for a and f, the most unsatisfactory ones from the standpoint of polarization.

One should keep in mind that the requirements imposed by Eq. (13) on the bandwidth are too severe, since in practice the frequency response of the receiving equipment will be closer to a Π-shaped curve. The primary cause of the depolarization expressed by Eq. (13) is the fact that the receiver frequency response will have prominent extended tails beyond the limits of the transmission band.

For the case of a Π-shaped frequency response, the depolarization in the band will be

$$m = 1 - \frac{\sin\Delta\varphi}{\Delta\varphi}, \tag{15}$$

where $\Delta\varphi$ has the same meaning as in Eq. (13).

For the same values of m, a, and f as in the first case, the bandwidth permitted by Eq. (15) would be about 600 kc. To increase the reliability of polarization measurements and to improve the resistance of the equipment to interference, it is desirable for all the reception channels of the spectrometer to have the smaller of these bandwidths, 60 kc.

Time Constant of the Output Filter. One important aspect of the design of receivers in radio astronomy is the choice of a time constant τ for the output filter. To enhance the sensitivity of the radiometer, τ should be increased. But it can only be increased to a certain limit, which differs for different equipment and problems. In our case, this limit is imposed mainly by the time required for a source to drift through the antenna beam. Since the DKR-1000 radio telescope is stationary in azimuth, the drift time of a source through the beam cannot be increased by means of guiding.

There is a noteworthy advantage here for multichannel spectral apparatus as compared to single-channel equipment, namely a gain in τ. The gain is achieved by the continuous and simultaneous analysis of many frequencies. In a single-channel system, different frequencies are analyzed successively, so that the time required for their analysis, and hence the quantity τ, is inversely proportional to the number of frequencies being analyzed.

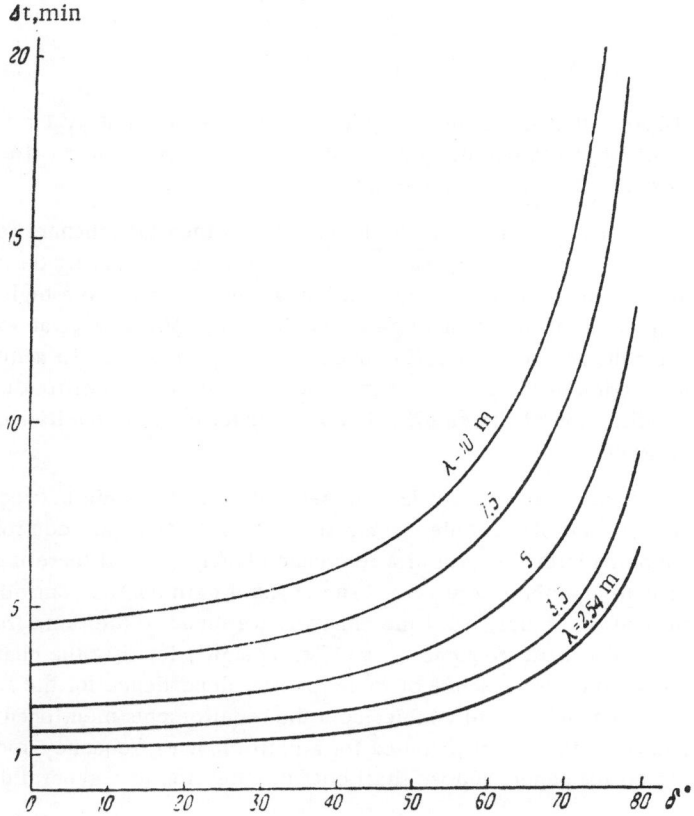

Fig. 10. Drift time of a source through the antenna pattern of the DKR-1000 as a function of declination of the source for selected wavelengths. The graph has been prepared for an antenna extending 500 m in an East−West direction.

Unlike amplitude distortions, the phase distortions introduced by the output filter transmit no error to the results of polarization measurements [24]. Hence, τ should be selected from the condition that the amplitude distortions of the filter be small.

Analysis of the passage of a low-frequency signal of period T through an output filter of the integrating-network type indicates that 10% amplitude distortions of the signal will occur if τ is equal to approximately 0.1 T. The transmission coefficient of the integrating network is

$$\frac{I_{out}}{I_{in}} = \sqrt{(R\,C\,\omega)^2 + 1},$$ (16)

where $\omega = 2\pi/T$ and T is the period of signal variation. If

$$\frac{I_{in}}{I_{out}} \leqslant 1.1,$$

then

$$\tau \leqslant 0.074\ T.$$ (17)

The period of signal variation at the output filter is determined by the time Δt required for a source to drift through the antenna beam between zero points. Here, Δt depends on the declination δ of the source and the wavelength. Figure 10 shows how Δt varies for several wavelengths in the spectrometer band, and for an antenna 500 m long:

$$\Delta t = \frac{\Delta q}{15 \cos \delta} = \frac{\lambda 0.456}{\cos \delta} \tag{18}$$

The curves of Fig. 10 show that the minimum value of Δt is approximately 1 min. Since we ordinarily have $T \approx 2\Delta t$, the time constant of the output filter satisfying Eq. (17) is about $\tau = 10$ sec for all the reception channels of a spectrometer operating with a 500-m antenna.

Matching of the Reception-Channel Gain. In the multichannel receiving system illustrated in Fig. 9, information on the spectrum or polarization is acquired after comparing the signals in different channels. In this method of analysis the gain of each channel should be the same to a high degree of accuracy. Under our conditions, the balancing of the channel gains should be performed over an extended time, equivalent to the observing time for the source of radiation under study (up to 24 h). To achieve this requirement by means of independent stabilization of the gain of each channel would be very difficult. In practice, there is no need for independent stabilization of the gains. Moreover, under certain conditions, large fluctuations may be tolerated in the channel gain.

The basic requirement is the need to stabilize the zero difference in gain between all channels. This requirement can be fulfilled by means of a simple slave system of automatic gain control (AGCA in Fig. 9). The gain of each channel is compared with the gain of a reference channel, and, if there is a difference, the automatic gain control reduces it to approximately zero. The channel gain can be controlled from the internal noise level of the apparatus and the signal, if all the channels are tuned to the same frequency. If the spectrometer channels are tuned to different frequencies, a differing signal level in the channels could arise not from a difference in gain, but from the presence of some spectral dependence for the signal. The frequency-modulation method in a multichannel system enables both the requirements mentioned to be achieved. A detailed analysis of the automatic gain control proposed for a multichannel frequency-modulation radiometer would fall beyond the scope of this paper, and we shall confine attention to a general descripton of the control principle.

If for half the modulation period all the spectrometer channels are tuned to the same "standard" frequency f_0, the gain of all channels can be controlled and compared with the signal and noise levels at the receiver output. The difference in level is determined in this case by the difference in channel gain, and can be utilized as an error signal for the automatic gain control. During the second half of the period, the reception channels are tuned to different frequencies and the automatic gain control is disconnected from the comparison circuit, although it continues to operate because of its own time constant. During this half of the modulation period, a signal determined by the spectral and polarization responses is recorded. A change in the value of the signal at the receiver input will not affect its gain with this control system, so that the control can be executed with a fully connected antenna pointed toward any part of the sky.

Semiautomatic Control of the Spectrometer Frequency Response. For absolute spectral polarization measurements with the spectrometer described here, a smooth frequency response over the entire frequency range under investigation is necessary for the radiometer, including the antenna. Under actual conditions, this requirement cannot be ensured over a wide frequency range. In our case, the chief irregularities in the frequency response arise primarily from the antenna system (see [10]).

In a multichannel frequency-radiometer system, this difficulty can be avoided by applying fine adjustment of the gain of individual channels to compensate for irregularity in the frequency response of the high-frequency channel.

If a signal with a smooth spectrum arrives at the spectrometer input, the signal observed at the spectrometer output will be proportional to the frequency response of the high-frequency channel. If this response is irregular, it can be used as an error signal for adjusting the gain in all channels. This correction to the gain should be applied to those circuit elements which have no effect on the transmission coefficient at the standard frequency f_0. The elements in question are exclusively the local oscillators of frequencies f_{1-14}.

The principle of the "frequency response control" is best understood from the operation of a single channel. If the transmission coefficients of the channel are different at the two modulated frequencies f_0 and f_1, then if a smooth spectrum is supplied to the channel input, a signal will be observed at its output which is proportional to the difference of the transmission coefficients at frequencies f_0 and f_1. This difference arises primarily from the frequency response of the high-frequency channel. The signal from the channel output can be used as an error signal for the automatic control system. The transmission coefficient of the channel at frequency f_0 cannot be changed, since it is associated with the automatic gain-control system described in the preceding section; thus, the control must be accomplished at frequency f_1. This can be done by controlling the amplitude of the f_1 local oscillator, and correspondingly the transmission coefficient of the transformer at frequency f_1. As a reference signal against which the frequency response of the spectrometer can be corrected, one may use the radiation from certain regions of the galaxy having a known spectrum, with a spectral index that is constant over the frequency range in question. With this type of calibration, the transmission coefficients of the spectrometer at frequencies f_{1-14} will represent the behavior of the calibration-signal spectrum, but irregularities in the frequency response of the spectrometer will be eliminated just as if a calibration signal with a smooth spectrum had been used.

When measurements are being made with the instrument, the automatic frequency-response control (AFC in Fig. 9) should be disconnected, since otherwise information would be lost on the spectrum of the source being observed. It should be switched on occasionally to correct for time drifts in the frequency response of the high-frequency spectrometer channel.

Literature Cited

1. V. A. Razin, Radiotekhn. i élektron. 1 : 846 (1956).
2. V. A. Razin, Astron. Zh. 35 : 241 (1958).
3. G. Westerhout, C. L. Seeger, W. N. Brown, and J. Tinbergen, Bull. Astron. Inst. Neth. 16 : 187 (1962).
4. R. Wielebinski, J. R. Shakeshaft, and I. I. K. Pauliny-Toth, Observatory 82 : 158 (1962).
5. R. X. McGee, O. B. Slee, and G. J. Stanley, Austral. J. Phys., 8 : 348 (1955).
6. I. M. Thomson, Nature, 180 : 84 (1957).
7. I. I. K. Pauliny-Toth, F. E. Baldwin, and J. R. Shakeshaft, Monthly Notices Roy. Astron. Soc. 122 : 21 (1961).
8. J. L. Pawsey and E. Harting, Australian J. Phys. 13 : 740 (1960).
9. V. V. Vitkevich, Vestn. Akad. Nauk SSSR, No. 5 : 23 (1961).
10. Yu. P. Ilyasov and A. D. Kuz'min, this volume, p. 7.
11. V. A. Udal'tsov, Radiofizika, 5 : 5 (1962).
12. V. A. Udal'tsov, Proceedings of the Fifth Conference on Problems of Cosmogony, Moscow, USSR Acad. Sci. Press (1956), p. 506.
13. F. Hatanaka, Publ. Astron. Soc. Japan, 7 : 114 (1955).
14. N. L. Kaidanovskii, É. G. Mirzabekyan, and S. É. Khaikin, Proceedings of the Fifth Conference on Problems of Cosmogony, Moscow, USSR Acad. Sci. Press (1956), p. 113
15. A. D. Kuz'min and V. A. Udal'tsov, Astron. Zh. 36 : 33 (1959).
16. V. A. Udal'tsov, A. D. Kuz'min, and A. E. Salomonovich, Astron. Zh. (in press).
17. V. A. Udal'tsov, Tr. Fiz. Inst. Akad. Nauk SSSR 17 : 169 (1962).
18. B. Y. Mills, Paris Symposium on Radio Astronomy, 1958, Stanford Univ. Press (1959) [Russian translation: Moscow, Foreign Lit. Press (1961), p. 422].
19. S. J. Goldstein, Proc. IRE 43 : 1663 (1958).
20. A. G. Kislyakov, Radiofizika 1 : 81 (1958).
21. B. M. Yanovskii, Terrestrial Magnetism, Moscow (1953).
22. J. V. Evans, Proc. Phys. Soc., 69 : 953 (1956).
23. J. V. Evans, Atmosph. Terr. Phys., 11 : 259 (1957).
24. V. V. Vitkevich, A. D. Kuz'min, L. I. Matveenko, R. L. Sorochenko, and V. A. Udal'tsov, Radiofiz. i Élektron. 9 : 1421 (1961).

SOME DISCRETE RADIO EMISSION SOURCES AT 9.6 cm

Yu. N. Vetukhnovskaya and A. D. Kuz'min

The first cycle of observations of discrete radio emission sources at 9.6 cm was made in January—April 1960 at the P. N. Lebedev Physics Institute of the Academy of Sciences of the USSR. The results of the observations are summarized in a catalog called "List A," which was published in [1].

These observations were continued in August—September 1961 and in February—May 1962. This paper describes the methods used in these observations and the results are given.

Apparatus and Observation Method

The observations were made with the 22-m radio telescope of the P. N. Lebedev Physics Institute of the Academy of Sciences of the USSR [2].

The electrical axis of the radio telescope was adjusted, using the discrete radio emission sources Cassiopeia A and Cygnus A.

A Dicke radiometer with fluctuation sensitivity 0.15°-0.20°K and a time constant of 24 sec was used. An additional horn radiator, situated in the focal plane and displaced 2.5λ relative to the main radiator, was used as the low-temperature equivalent in the observations of 1961. Such an equivalent receives sky radio emission in a direction close to the electrical axis of the radio telescope, which makes difficult investigation of elongated radio emission sources. For this reason, the equivalent used in the observations of 1962 was an auxiliary antenna, a horn with an aperture $5\lambda \times 6\lambda$, set in such a way that its electrical axis was 30° higher than the electrical axis of the radio telescope.

Intensity of the signal received was measured with a gas-discharge noise generator. The latter was calibrated, using the discrete radio emission sources Taurus A and Cassiopeia A, whose radio emission flux density was assumed to be $790 \cdot 10^{-26}$ and $1450 \cdot 10^{-26}$ W $\cdot$ m^{-2}cps^{-1}, respectively.

Two orthogonal profiles were made for all sources with declinations $\delta < 40°$: for right ascension α (with a fixed antenna due to the earth's diurnal rotation) and for declination δ (α was tracked with a change of δ, introduced into the instrument for transforming coordinates).

Usually only one α profile was made for sources with $\delta > 40°$; in such cases, δ was assumed known on the basis of data from other radio emission observations. However, for some sources (Nos. 4, 7, and others), several α profiles were made with different δ settings.

Antenna temperature, transit time, and width of the transit curve were directly measured parameters.

Since the radiometer time constant $\tau = 24$ sec is commensurable with the time of transit of the source through the radio telescope antenna directional diagram, a correction for the time constant was made when determining the parameters mentioned. The correction was computed using the curves given in [1], which we first checked experimentally at several points by a comparison of the records of the strong discrete sources Omega, Orion, Virgo, Taurus, Cygnus, and Cassiopeia, made with time constants of 4 and 24 sec.

The angular dimensions θ, flux density S, and brightness temperature T_{br} of the sources were computed from known relationships, as given in the following set of tables (see, for example, [1]):

YU. N. VETUKHNOVSKAYA AND A. D. KUZ'MIN

Table 1 (List B)

No.	Right ascension			Declination			Galactic coordinates				Angular dimen. θ
	$\alpha_{1950,0}$	$\pm\Delta\alpha$	P_α	$\delta_{1950,0}$	$\pm\Delta\delta$	P_δ	l^{I}	b^{I}	l^{II}	b^{II}	
1	00h40m08s	18s	3s.37	[51°47′]	—	0′.33	90°	—10°2	123°	—10°5	[<1′][7]
2	01 06 10	10	3.16	13 02	6′	0.32	100	—48.4	133	—48.9	[<1′][7]
3	01 34 38	7	3.42	32 51	3	0.30	102	—28	134.3	—28.5	[<1′][7]
4	02 02 03	5	4.50	64 35	3	0.30	98.5	4	131.5	3.5	[5′][10]
5	02 19 54	20	3.78	[42 45]	—	1.27	108	—16	141	—16.8	[5′±2′][6]
6	03 23 07	5	4.56	[55 07]	—	0.21	112	0	145	—0.8	[<3′][7]
7	03 59 49	17	4.50	51 10	1	0.17	118.3	0	151.4	—1	16±3
8	4 15 12	6	4.01	[37 51]	—	0.15	130°	8°	163°	6.8	[3′][7]
9	5 01 15	15	4.08	[38 02]	—	0.09	135.5	—0.3	163°	—0.2	[<1′][7]
10	5 14 55	24	3.95	[33 48]	—	0.07	140.8	—0.4	174.3	—0 3	33′±4′
11	05 38 54	12	4.65	[49 51]	—	0.03	129.3	12 3	164.2	12.4	[<0′.2][7]
12	06 24 52	8	2.93	[—5 51]	—	0.04	183.5	—6	216.3	—6	[<1′][7]
13	08 09 58	5	4.34	[48°23′]	—	0.18	08.3	33.8	173	32.8	[<0′.2][7]
14	12 16 47	5	3.06	6 01	7	—0.33	253	68	286.5	68	[55″][17]

Antenna temp., T_A, °K	Flux dens. $S \cdot 10^{26}$ W·m^{-2}·cps^{-1}	Brightness temp.	Spectral index	Identification opt.	Identification radio	Notes
0.47±0.05*	6.8±0.7*	—	0.55±0.10 [0.61±0.11][8]	—	HR 5, WB 6, ERL, 3C20, 2C59	
0.41±0.05*	6.0±0.7*		0.65±0.10 [0.63±0.07][8] [0.65±0.06][32]		3C33, HR10, ERL, 2C94, Bl8, HM, Bl, Wh 4	
0.53±0.05*	7.7±0.7*	—	0.55±0.10 [0.46±0.10][8] [0.52±0.09][32]	—	HR 15, 3C 48, ERL, G, 2C 133	
2.25±0.15	35±2.5	50.7	T	---	3C 58, HR 16, 2C 177, WB 8, AMSWW 6	$ME=1.32 \cdot 10^5$ 1800 ps$< r_{neb} <$4200 ps $1.3 \cdot 10^3 < \frac{M}{M_\odot} <$ $< 1.1 \cdot 10^4$ $70 < N_e < 110$ $12 < 2s < 26$
0.45±0.20	7.0±2	—	0.55±0.05 [0.44±0.14][8] [0.54±0.07][32]	—	3C 66, HR 18, HM, RSE, 2C 205	
0.54±0.20	7.8±3	—	0.40±0.10 [0.49±0.10][8]	—	3C 86, ERL, HR24	
1.1±0.20	27.2±6	3.3	T	NGC 1491 (ШГ 36)	WB 12	$A_v=2.82$, $a_v=3.6$ $ME=8.6 \cdot 10^3$, $r_{enb}=$ $=800$ ps, $\frac{M}{M_\odot} \cong 50$, $N_e = 75$cm^{-3} $2s = 1.5$ps
0.61±0.08*	9.15±1.0	—	0.65±0.10 [0.66±0.09][8] [0.8±0.4][14]	—	3C 111, ERL, HR 30, B 2 Bl 20, 2C 379, L 9, Wh 12, HW 9, C, SWC	
0.20±0.08*	3±1*	—	0.95±0.05 [0.90±0.10][8] [1.2±0.20][14]	—	3C 134, ERL, HR 34, WB 15, 2C 440, HW 10, C, Bl 23, Wh 14 RSE, SWC	
0.55±0.15	32±9	1.1	T	Emission nebula IC 405	WB 16, L 10	Exciting star 34078 $ME=2.8 \cdot 10^3$ $r_T=530$ ps , $r_*=$ $=500$ ps, $N_e=16.6$, $A_v=1^m.56$
0.82±0.08	11.9±1	—	0.50±0.15 [0.49±0.07][8]	—	3C 147, ERL, HR 39, 2C 485, Wh 16, HW 12	
0.77±0.05*	11.1±0.7*	—	0.55±0.10 [0.60±0.09][8]	—	3C 161, ERL, HR 42, 2C 553, MSH 06—04, HM, KH, SWC	
0.48±0.05*	7.0±0.7*	—	0.70±0.10 [0.72±0.09][8] [0.85±0.2][14] [0.73±0.09][32]	—	3C 196, ERL, HR 45, 2C 724, C, HM, HW 31, E, RSE, SWC, Wh 22	
0.76±0.15	11.0±2	—	0.57±0.20 [0.19±0.19][8] [0.24±0.15][32]	—	3C 270, HR 52 MSH 12+05, HM, KH	Acc. to [17], normal elliptical galaxy

Table 1 (List B) (continued)

No.	Right ascension			Declination			Galactic coordinates				Angular dimen. θ
	$\alpha_{1950.0}$	$\pm\Delta\alpha$	P_α	$\delta_{1950.0}$	$\pm\Delta\delta$	P_δ	l^I	b^I	l^{II}	b^{II}	
15	13h28m48s	10s	2s.82	[30°40']	—	—0'.31	21°	79°	62.5	80.2	[3'±1']^6
16	4 09 22	15	2.14	[52 25]	—	—0.28	63	61	98.5	61.4	[4".5]^28
17	18 28 16	5	1.56	[48 41]	—	0.04	44.8	22.6	79.4	23.4	[<1']^7
18	19 01 26	10	2.95	5 22	8	0.09	7.3	—2.1	40.4	—0.7	0°.50± ±0°.05
19	19 04 55	10	2.91	6 59	7	0.09	9.1	—2.2	42.4	—0.8	[9']^10
20	19 22 41	5	2.70	16 00	3	0.12	19.1	—1.6	52.2	—0.3	0'.46±0'.1
21	19 59 25	5	2.31	33 21	4	0.16	38.1	0.6	71.2	1.5	18'·24'±3'
22	20 12 34	5	2.59	[23 25]	—	0.18	31	—7	64	—6	[<1'.5]^7
23	22h43m36s	13s	2s.72	[39°25']	—	0'.32	0°.7	—8°.1	99°.8	—7°.8	[4'±0'.5]^7

Notes: 1) Notations for column 18: 3C [6]; ERL [7]; HR [8]; L [18]; B [19]; Bl [20]; HW [21]; C [22]; E [23]; RSE [24]; SWC [25]; 2C [9]; WB [10]; W [11]; HM [12]; MSH [13]; WH [15]; KH [16]; LMH [26]; HMS [27]; G [29]; AMSWW [30]; Mü [31]. 2) In columns 12 and 16 the superscripts are references to the literature. 3) The asterisk denotes the results of control observations with a receiver having a fluctuation sensitity of 0.5°K.

$$\theta = \sqrt{\varphi^2_{obs} - \varphi^2_{ant}}, \tag{1}$$

$$S = g \frac{2kT_A}{A}, \tag{2}$$

$$T_{br} = \frac{1}{1-\beta} hT_A. \tag{3}$$

Here it was assumed that the radio brightness distribution for the source and the antenna directional diagram are approximated by Gaussian curves.

Antenna temp. T_A, °K	Flux dens. $S \cdot 10^{26}$ W·m^{-2}·cps^{-1}	Brightness temp.	Spectral index	Identification		Notes
				opt.	radio	
0.72±0.07	10.5±1	—	0.38±0.10 [0.12±0.09]^8	—	3C 286, HR 60, 2C 1120, HW 92	
0.79±0.07*	11.4±1*		0.57±0.10 [0.53±0.10]^8 [0.60±0.20]^{14}	—	3C 295, HR 62, ERL, 2C 1175, SWC, C, HM, HW 101, Wh 37, AMSWW, IAU 14 N$_5$A	Acc. to [28] galaxy with $v=$ $=140\,000$ km/sec
0.57±0.15	8.3±2	—	0.72±0.05 [0.71±0.07]^8 [1±0.4]^{14}	—	3C 380, ERL, HR 79, 2C 1569, RSE, B 20, SWC	
1.22±0.20	61±11	2,5	T	—	3C 396, WB 65, LMH 39	$ME=6.5 \cdot 10^3$, $r_T \geqslant 800$ ps, $\frac{M}{M_\odot}>$ >200, $N_e < 28$ cm^{-2} $2s > 8$ ps
1.50±0.15	26.5±5	9,8	T	—	3C 397, WB 67, LMH 42	$ME=2.6 \cdot 10^1$, $r \geqslant$ $\geqslant 1100$ ps $\frac{M}{M_\odot}>50$, $N_e < 90$ cm^{-3} $2s > 3$ ps
1.70±0.15	76±7	3,7	T	—	WB 74, LMH 49	$ME=9.6 \cdot 10^3$ $r_T > 1900$ ps, $\frac{M}{M_\odot}>$ $>1.9 \cdot 10^3$, $N_e < 100$, $2s > 1$ ps
1.25±0.20	40±5	3,3	T	S204	WB 82, W 58, Mü 31	$ME=8.7 \cdot 10^3$, $A_v=5^m.5$, $r=$ $=200$ ps, $N_e=60$, $\overline{M_\odot}=160$, $2s=$ $=2.4$ ps
0.34±0.05*	4.9±0.7*	—	0.85±0.10 [0.87±0.08]^8 [1.1±0.3]^{14}	—	HR 89, 3C 409, ERL, SWC, 2C 1686	
0.55±0.15	8.3±2	—	0.70±0.10 [0.83±0.09]^8 [1.7±0.5]^{14}	—	3C 452, ERL, HR 103, 2C 1870	

Results of Observations

The results of the observations are given in the table, which we have called "List B." The observed sources are arranged in the sequence of increase of right ascensions. Column 1 gives the sequence number of List B. Columns 2, 4, 5, 7 give the coordinates of the sources, related to the epoch 1950.0 and precessions for α and δ. Columns 8, 9, 10, 11 give the galactic coordinates in the old (l^I, b^I) and new (l^{II}, b^{II}) systems. Column 12 gives the angular dimensions of the sources, determined from the broadening of the transit curve or taken from other sources. Columns 13, 14, 15 give antenna temperatures, flux densities, and brightness temperatures. The mean square errors of determination are given for all the mentioned parameters; these correspond to a fluctuation sensitivity of 0.2°K.

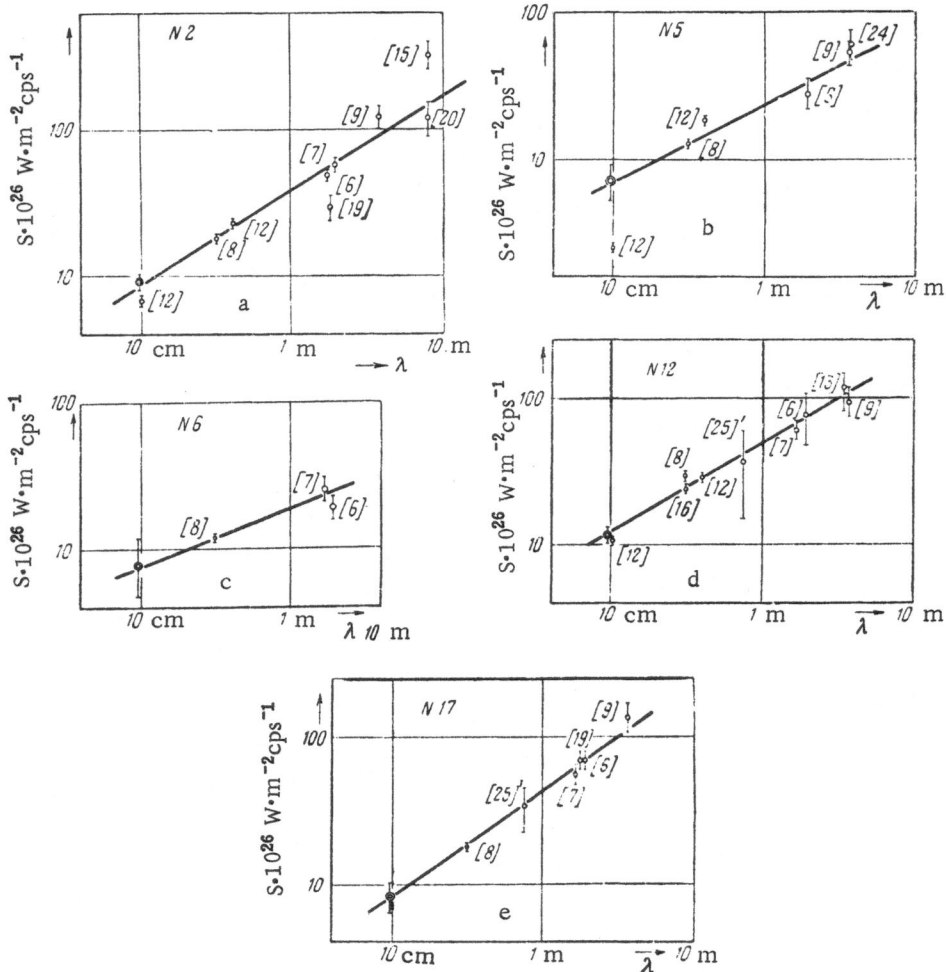

Fig. 1. Spectra of some nonthermal radio emission sources.

A comparison with the results of other investigators [6-33] was made and the radio source spectra were determined. In some cases a correction of flux densities cited by these investigators was made, such as was done in [1]. In addition, if the angular dimensions of the sources which we obtained differed from the corresponding values given in [6-33], the flux densities cited there were recalculated to our angular dimensions (for example, for No. 18).

The observed sources were classified as thermal and nonthermal on the basis of the spectral type. The spectra of the nonthermal sources (some of which are shown in Fig. 1) were used to determine their spectral indices, given in column 16 of the table.

Comparison of the determined spectral indices with the data of Harris and Roberts [8] does not reveal any systematic difference in the results. However, it should be pointed out that the spectral index n = − 0.19 ± 0.19 for NGC 4261 (No. 14) cited in [8] is unrealistic because the value used for the flux density from the 3C catalog obviously is too low in this case. Using five points [12, 13, 16] we found n = 0.57. The spectral index of source No. 15 is higher (n = 0.38) when using the results of our measurements than is given by Harris and Roberts (n = 0.12 ± 0.09). The spectral indices for the other sources agree within the limits of error.

The spectral indices obtained by Whitfield [14] appear too low in some cases, such as for No. 23, but for the most part the differences are small.

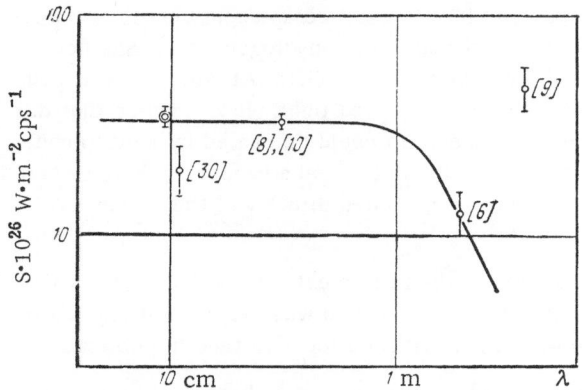

Fig. 2. Spectrum of the radio emission source No. 4.

Yu. P. Pskovskii [32] determined the differential spectral indices of some strong radio emission sources, using the most reliable results obtained in 35 studies. His data coincide almost completely with the results of our study.

Sources of Thermal Radio Emission. Seven of the discrete radio emission sources which we observed have a thermal character. The spectra of some of them are shown in Fig. 2.

In order to identify the radio emission sources with emission nebulae, we determined the exciting stars and reckoned the distances, linear dimensions, electron density, and mass [4]; for radio emission sources not identified with optical objects we made approximate estimates of these characteristics.

The emission measure was determined directly from observational data on the assumption of a homogeneous isothermal nebula with the electron temperature $T_e = 10^4$ °K:

$$ME = 2.4 \cdot 10^5 \frac{T_{br}}{\lambda^2}.$$

The determined emission measures are given in column 19.

One of the interesting objects which we observed is source No. 10. It is identified with the emission nebula IC 405. According to [35, 36], the exciting star for this nebula is the variable AE Aur, Sp O9.5 v, which varies in brightness from $5^m.4$ to $6^m.1$.

However, this conclusion was drawn in [35, 36] solely on the basis of the coincidence of the star and the nebula in the figure plane. In order to check the assertion that the star AE Aur actually excites the nebula IC 405, we will consider whether they coincide for a third coordinate as well. This will be done by determining the distance to the star and to the nebula independently.

The absorption to the nebula, using radio emission data, is determined as

$$A_v = 1.2 \left(m_{n_\alpha} - 4.5 + 2.5 \log \frac{T_{br}}{\lambda^2} \right). \tag{4}$$

With absorption determined in this way taken into account, the distance to the star AE Aur is 500 parsecs (Johnson [46] gives 580 parsecs).

Distance to a nebula with a known exciting star can be determined using (4)

$$r_{neb} = 21.3 \, u^{3/2} \, (S \cdot 10^{26})^{-1/2}, \tag{5}$$

where u is the radius of the ionization zone of the exciting star or group of stars when $N_e = 1$ cm^{-3}. According to Ikhsanov [45], for a star of the spectral class O9.5, u = 26.5; the distance to the nebula, therefore, is 530 parsecs.

The distance to the star and to the nebula is virtually identical. The nebula and star therefore apparently actually occupy the same space and the mentioned star may be the exciting star. In this case, known relations (4) can be used to determine the mass, electron density, and dimensions of the nebula. They are given in column 19 of the table.

IC 405 is a filamentary nebula with an intense and irregular system of filaments at the center and a fainter system at the margins. In the computations of mass, density, and dimensions, we used a model of a spherically symmetrical nebula and due to the inaccuracy of this approximation there can be considerable discrepancies with the same results obtained from optical observations [38].

As early as 1939, Henyey and Greenstein [43] pointed out that IC 405 has a conspicuous continuous spectrum. Gaze [44] obtained photographs of the nebula in integrated light and in the hydrogen lines. She notes that the pictures are completely different. A rather bright arc, emanating from the star AE Aur, which is completely invisible on the photographs in the hydrogen lines, is visible in integrated light. It is assumed that this arc also is responsible for the continuous spectrum. The continuous spectrum could be caused by a dust nebula, but it is improbable that a dust cloud is present so close to the hot star Sp O9.5. Ambartsumyan [47] feels that the arc apparently is a cometary inclusion (of the cometary nebula type), related directly to the variable AE Aur and possibly is a stream flowing from the star.

On the basis of radio emission data, due to the low resolution of the instruments, it is difficult to establish what contribution to the radio emission is made by the cometary inclusion and whether there is any radio emission at all from objects of the cometary nebula type (most of them with too low a surface brightness).

In addition, as mentioned above, the nebula IC 405 is excited by a variable. No changes of brightness of the nebula were observed in the optical region. It therefore is of great interest to attempt detection of changes of flux density at radio frequencies and to obtain the detailed radio brightness distribution for the nebula IC 405.

Source No. 7 is identified with the emission nebula NGC 1491. The form of the records which were obtained for several declinations with transit through α indicates a possibility of asymmetry of the radio source.

Hubble, in [37], gives the star Sp BO, m = 10.6, as the star exciting NGC 1491. His hypothesis can be checked by determining the distance to a nebula with a measured flux density and an Sp BO exciting star by the known method [4]. The distance to the nebula is 160 parsecs. The distance to the star, taking into account total absorption, determined using radio emission data and formula (1), is 3800 parsecs. Therefore, there is no basis for assuming that Hubble's star is the exciting star for nebula NGC 1491. Sharpless believes that the nebula is excited by the star BD 50°886, but gives neither the spectral class nor the distance to the star. No other stars of early spectral classes were found here.

The radio source No. 21 is identified with the emission nebula S 204, $m_{H_\alpha} = 12^m.8$ [35]. According to [38], it apparently is excited by clusters of stars of early spectral classes, investigated by Numerova [39], such as B1, No. 42. He gives the distance to them as 1900 parsecs. The distance to the nebula, determined from the ratio of total [formula (4)] and differential absorptions [42], is 2000 parsecs. In this case, $N_e = 6$, $M/M_\odot = 160$. This agrees well with the results obtained by Pariiskii [34], obtained using Westerhout's data $r_{neb} = 2100$ parsecs, $N_e = 20$, dimension 26 parsecs, and $M/M_\odot = 150$.

Optical analogs were not found for the four thermal sources (Nos. 4, 18, 19, 20). However, it is known that objects with a surface brightness m > 14^m cannot be observed from the earth due to the existing background of the night sky. We, therefore, postulate that the radio sources Nos. 4, 18, 19, and 20 nevertheless are H II regions which are invisible in the telescope due to the great absorption in the interstellar medium. We will attempt to estimate the possible distance to them, and also the mass, density, and dimensions. Assuming that $m_{H_\alpha} > 14^m$, and using a spherically symmetrical homogeneous model, (4) can be used to determine the lower absorption limit. Then, assuming the differential absorption to be known (for example, see [42]), we find that $r_{neb} \geq A_V/a_V$.

The possible lower limit of mass is determined using the formula

$$\frac{M}{M_0} = 3.94 \cdot 10^{-6} \Theta^{3/2} r^{5/2} (S \cdot 10^{26})^{1/2},$$

where Θ is the dimension of the source in degrees, S is flux density.

It also is possible to estimate electron density

$$N_e^{-5} = \frac{0.175 M E^3}{2^{3.75} M/M_\odot}$$

and dimensions

$$2S = \frac{ME}{N_e^2}.$$

For sources not lying in the plane of the galactic equator, it is possible to make at least a rough estimate of the possible upper limit of distance. It apparently can be assumed that HII regions form part of the plane subsystem of the galaxy with a thickness of about 250 parsecs. The distance to these regions therefore should not exceed

$$r_{max} = \frac{250}{\sin b^{II}}.$$

Among the above-mentioned sources, only one (No. 4) has a galactic latitude $b^{II} \neq 0$. The upper limit of distance to it, determined using relation (3), was 4200 parsecs. Thus, for source No. 4, 1800 parsecs $< r_{neb} <$ 4200 parsecs. The only possible exciting star for a nebula situated at such a distance is a star of a spectral class earlier than O7. With the determined limits of distances, a star of class O7, with the determined absorption $A_v = 6^m.3$ taken into account, has an apparent magnitude $10^m.5 < m_v < 13^m.5$, that is, to all intents and purposes it should be detectable. The only hot star which is close in coordinates, HD 12882 (Sp B61a, $M_v = -6^m.9$, $m_v = 7^m.5$, $r = 3600$ parsecs [40]) cannot be the exciting star because the maximum admissible distance in the figure plane, according to Ikhsanov [45], under such conditions should not exceed 2' and HD 12882 is situated at a distance of 42' from the center of the radio emission source. The absence of an exciting star can suggest some different mechanism of excitation of the nebula or that the radio emission source is not related to a HII region.

Only estimates of the lower limits of the distances to sources Nos. 18, 19, and 20 were made. The estimates of distances, masses, densities, and dimensions are given in column 19 of the table. The determined values of masses, densities, and dimensions are fully applicable for diffuse nebulae. Thus, the results are not contradicted by the assumptions made concerning the possibility of the existence of optically unobserved clouds of interstellar hydrogen in the region of the radio emission sources.

In the case of source No. 4, the electron temperature can be estimated from the inflection of the spectral curve

$$T_e = 10^4 \left(\frac{\lambda_n}{1.44}\right)^{1,5} \left(\frac{ME}{10^5}\right)^{0,72}.$$

According to Fig. 2, $\lambda_n = 1.15$ m; hence, $T_e = 8700°K$.

The authors are sincerely grateful to G. G. Basistov, N. F. Il'in, A. N. Kozlov, V. N. Kolyade, V. N. Koshchenko, L. A. Levchenko, and M. T. Levchenko for preparing the apparatus and assistance in observations.

Literature Cited

1. A. D. Kuz'min, Tr. Fiz. Inst. Akad. Nauk SSSR, 17:84 (1962).
2. P. D. Kalachev and A. E. Salomonovich, Radiotekhn. i Élektron., 8:422 (1961).
3. A. M. Karachun, A. D. Kuz'min, and A. E. Salomonovich, Radiotekhn. i Élektron., 6(3):430 (1961).
4. A. D. Kuz'min and R. I. Noskova, Astr. Zh., 39(2):241 (1962).
5. A. D. Kuz'min, Astr. Zh., 39(1):22 (1962).
6. D. O. Edge, J. R. Shakeshaft, W. B. McAdam, J. E. Baldwin, and S. Archer, Mem. Roy. Astr. Soc., 68:37 (1959).
7. B. Elsmore, M. Ryle, and Patricia R. R. Leslie, Mem. Roy. Astr. Soc., 68:61 (1959).
8. D. E. Harris and J. A. Roberts, Publ. Astron. Soc. Pacific, 72(427):237 (1960).
9. J. R. Shakeshaft, M. Ryle, J. E. Baldwin, B. Elsmore, and J. Thompson, Mem. Roy. Astron. Soc., 67:106 (1955).
10. R. W. Wilson and J. G. Bolton, Publ. Astron. Soc. Pacific, 72(428) (1960).
11. G. Westerhout, Bull. Astron. Inst. Neth., 14(488) (1958).

12. D. S. Heeschen and B. L. Meredith, Publ. NRAO, 1(8) (1961).

13. B. J. Mills, O. B. Slee, and E. R. Hill, Australian J. Phys., 11 : 360 (1958).

14. G. R. Whitfield, Monthly Notices Roy. Astron. Soc., 117 : 680 (1957).

15. G. R. Whitfield, Monthly Notices Roy. Astron. Soc., 120 :581 (1960).

16. K. J. Kellermann and D. E. Harris, Observ. Calif. Inst. Technol., 7 (1960).

17. B. J. Mills, Australian J. Phys., 13 : 550 (1960).

18. C. R. Lynds, Publ. NRAO, 1(3) (1961).

19. A. Boischot, Radioastronomy, Paris Symposium, 1958, paper 89.

20. J. H. Blythe, Monthly Notices Roy. Astron. Soc., 120 : 581 (1960).

21. C. Hazard and D. Walsh, Astron. Contributions Univ. Manchester, 1(6) : 338-350 (1960).

22. R. G. Conway, Monthly Notices Roy. Astron. Soc., 117 : 692 (1957).

23. B. Elsmore, Monthly Notices Roy. Astron. Soc., 118 : 603 (1958).

24. M. Ryle, Smith, and B. Elsmore, Monthly Notices Roy. Astron. Soc., 110 : 508 (1950).

25. C. L. Seeger, G. Westerhout, and R. G. Conway, Astrophys. J., 126 : 585 (1957).

26. M. J. Large, D. S. Mathewson, and C. G. T. Haslam, Monthly Notices Roy. Astron. Soc., 123 : 113 (1961).

27. P. T. Haddock, C. H. Mayer, and R. M. Sloanaker, Astrophys. J., 119 : 456 (1954).

28. L. R. Allen, H. P. Palmer, and B. Rowson, Nature, 188(4752) : 731 (1960).

29. S. J. Goldstein, Astron. J., 67(3) : 171 (1962).

30. W. A. Altenhoff, P. S. Mezger, H. Strasst, H. Wender, and G. Esterhout, Veröff. der Universitäts-Sternwarte zu Bonn, No. 59 (1960).

31. H. G. Müller, Veröff. der Universitäts-Sternwarte zu Bonn, No. 52 (1959).

32. Yu. P. Pskovskii, Astr. Zh., 39(2) : 222 (1962).

33. H. J. Smith and Dorrit Hofflein, Publ. Astron. Soc. Pacific, 73(434) (1961).

34. Yu. N. Pariiskii, Izv. Glavnaya Astrofiz. Observ. (Pulkovo), 21 : 164, 5 (1960).

35. V. F. Gaze and G. A. Shain, Izv. Krymsk. Astrofiz. Observ., 15 : 11 (1955).

36. S. Sharpless, Astrophys. J., Supplement No. 41 (1959).

37. E. P. Hubble, Astrophys. J., 56 : 162 (1922).

38. R. E. Gershberg and L. P. Metik, Izv. Krymsk. Astrofiz. Observ., 24 : 148 (1960).

39. A. B. Numerova, Izv. Krymsk. Astrofiz. Observ., 19 : 189 (1958).

40. W. A. Hiltner, Astrophys. J., Supplement No. 24 (1956).

41. G. de Vaucouleurs, Astrophys. J., Supplement No. 56 (1961).

42. P. P. Perenelo, Astr. Zh., 22 : 129 (1945).

43. G. Greenstein and L. Henyey, Astrophys. J., 89 : 653 (1939).

44. V. F. Gaze, Izv. Krymsk. Astrofiz. Observ., 10 : 213 (1953).

45. R. N. Ikhsanov, Izv. Krymsk. Astrofiz. Observ., 23 : 31 (1960).

46. H. M. Johnson, Astrophys. J., 118 : 370 (1953).

47. V. A. Ambartsumyan, Works, Vol. 2.

INVESTIGATION OF LARGE-SCALE IONOSPHERIC INHOMOGENEITIES BY RADIOASTRONOMICAL METHODS

Yu. L. Kokurin

It was discovered in early investigations [1-3 et al.] that the direction of arrival of radio waves from any extraterrestrial source at an observation point on the earth's surface is subject to slow irregular changes. The value of these angular oscillations was proportional to the square of wavelength [4], which gave basis for postulating that they are caused by the regular refraction of radio waves in electron inhomogeneities of the earth's atmosphere. This paper gives the results of investigations of this phenomenon by the radioastronomical method.

Measurement of the Linear Dimensions of Inhomogeneities and Their Height Above the Earth's Surface from Observations of Vertical Ionospheric Refraction Near the Horizon

Refraction of radio waves in the earth's atmosphere is defined as the angle between the true and apparent directions to the cosmic source emitting radio waves. The true position of the source is determined from astronomical data. Refraction measurements therefore involve measurement of the angle of wave incidence.

In 1951, we made the first measurements of oscillations of the zenith angle of arrival of radio waves from two cosmic sources [5]. The measurements were made using a vertical radio interferometer, making possible observations of sources at the time of their ascension, i.e., in directions close to the horizon. The vertical or "marine" interferometer consists of a single antenna erected on an elevation above the sea surface (Fig.1). The interference diagram is formed by the superposing of two waves — direct, and reflected from the sea surface. The diagram is easily computed geometrically and, in particular, the positions of the interference maxima and minima are determined by the relations

$$z_m = \psi_m + \theta_m;$$

$$\left.\begin{array}{l} \dfrac{m\lambda}{2} = 2\left[(r_0 + H)\cos\theta_m - r_0\right]\cos\psi_m; \\[2mm] \cos\psi_m = \left[1 + \left(\dfrac{\sin\theta_m}{\cos\theta_m - \dfrac{r_0}{r_0 + H}}\right)^2\right]^{-1/2}. \end{array}\right\} \tag{1}$$

Here, z is the zenith angle; m is a whole positive number, even m corresponds to interference minima, odd m corresponds to interference maxima; H is antenna elevation above sea level; ψ is the angle of wave incidence on the sea surface; θ is the angle between directions from the center of the earth to the antenna and reflection point; r_0 is the earth's radius.

Vertical refraction is defined as

$$R(z_m) = z_m^{(a)} - z_m,$$

129

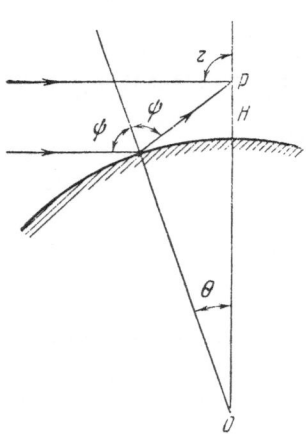

Fig. 1. Path of rays in vertical
(marine) interferometer.

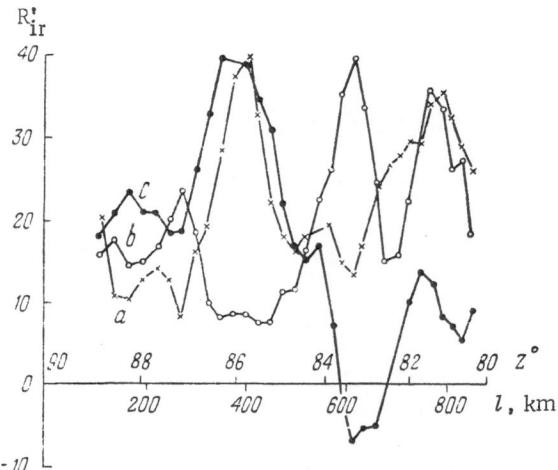

Fig. 2. Curves of irregular ionospheric refraction.
a) Taurus, June 6, 1951; b) Taurus, July 1, 1951;
c) Virgo, August 12, 1951.

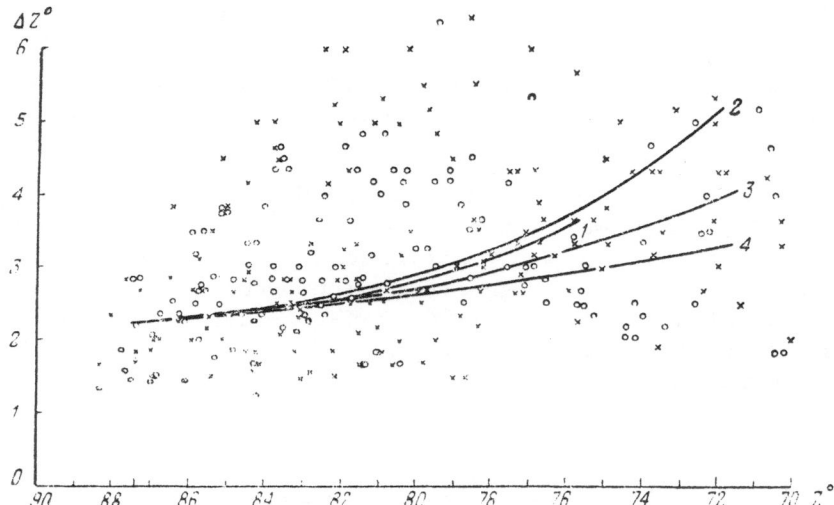

Fig. 3. Dependence of angular dimensions of inhomogeneities on zenith
angle. 1) Experimental curve; 2, 3, 4) computed curves for altitudes of
350, 800, and 2000 km.

where $z_m^{(a)}$ is the apparent zenith angle of the source; z_m is the true zenith angle of the source at the same
time.

The radio emission sources used in this series of measurements were the discrete sources Taurus ($\alpha =$
$05^h 31^m$, $\delta = +22°04'$) and Virgo ($\alpha = 12^h 28^m$, $\delta = +12°44'$). Measurements were made at $\lambda = 4$ m. Reception
was with a cophasal array with a diagram aperture of ~20° in the horizontal plane and ~40° in the vertical
plane. The antenna was erected at an elevation of 421.86 m above sea level on Mount Kastel' (Crimea, latitude 44°N) in such a way that the diagram axis was directed horizontally. The antenna was manually oriented
in the azimuth of the source. The open sea horizon permitted observations of each of the sources beginning
immediately with their ascension and to zenith angle ~70°.

Ninety-five observations of Taurus and 60 observations of Virgo were made from June through October 1951. Curves of the dependence $R = f(z_m)$ were constructed for these observations. In 111 cases these curves had a monotonic character, corresponding to regular refraction, and these will not be considered here. In the remaining 44 cases, the curves had a nonmonotonic, oscillating shape. Figure 2 shows three such curves. These curves make it possible to estimate only the angular dimensions of the inhomogeneities and the irregular refraction which they cause. The value of the irregular refraction in some cases attains $R_{ir} \sim 20'$, but an average for all observations is $\overline{R}_{ir} = \pm 12'$. In order to determine the linear dimensions of the inhomogeneities and the horizontal electron gradients of the ionospheric layer, it is necessary to measure their height above the earth's surface. In estimating this height, we utilize the fact that the law of change of angular dimensions of inhomogeneities with zenith angle is dependent on their height above the earth.

Figure 3 shows the apparent angular dimensions Δz of inhomogeneities, determined as the angular distances between succeeding maxima or minima of the refraction curves on the assumption that the inhomogeneities are motionless. It can be seen that there is a tendency to an increase of the angular dimensions Δz with a decrease of z. The figure also shows that the experimental values Δz have an extremely great scatter, which is related to the scatter of the linear dimensions of inhomogeneities and their motion in the layer. The scatter of linear dimensions of inhomogeneities cannot be estimated directly.

Curve 1 in Fig. 3 was constructed using experimental points on the assumption that, for each value of z, the distribution of values Δz is due solely to the movement of inhomogeneities (the type of ionospheric winds). Numerous experimental data on velocities of ionospheric winds were used in the computations. The distribution of directions of movement was assumed to be isometric in the plane of the layer.

Figure 3 also shows $\Delta z = f(z)$ curves computed for different heights h for a plane model of inhomogeneities (thin disk lying in the plane of the layer). All curves were normalized using the experimental value Δz for $z = 87°30'$.

The experimental curve of the variation $\Delta z = f(z)$ corresponds to a height $h \sim 350$ km. If this height is used, the horizontal linear dimensions of inhomogeneities are ~ 200 km.

Dependence of Irregular Ionospheric Refraction on Zenith Angle and Its Relationship to a Model of Ionospheric Inhomogeneities

Two possible models of ionospheric inhomogeneities can be proposed: model 1 (Fig. 4a) — a spherical ionospheric layer L, containing irregular horizontal gradients with the optical thickness $\int Ndh$, and model 2 (Fig. 4b) — an undulating layer without horizontal $\int Ndh$ gradients. With an appropriate selection of the parameters (grad $\int Ndh$ or grad h), any of these models can satisfy the experimental data. In particular, the above-cited value $\overline{R}_{ir} = \pm 12'$ for $z = 70-90°$ corresponds to the following parameters of the inhomogeneities: for model 1, the value $\int Ndh$ varies along the layer by 2-3%; for model 2, the value $\Delta H \sim 0.35$ km for a layer with a thickness ~ 100 km.

A selection can be made between these two models by an investigation of the dependence of irregular refraction $\overline{R}_{ir}$ on zenith angle z, which is different for these models. This dependence cannot be determined on the basis of the above-cited experimental data, due to the limited range of zenith angles ($z = 70-90°$) investigated.

Our earlier paper [6] contains approximate computations of the dependence of the mean amplitude of oscillations of refraction $\overline{R}_{ir}$ on zenith angle for both models on the assumption that the layer is thin (i.e., the vertical dimension of inhomogeneities is considerably less than the horizontal dimension) and that the horizontal gradients change along the layer in conformity to the sinusoidal law. Figure 5 gives the results of the computations (curve 1, for model 1; curve 2, for model 2). The curves were normalized using the experimental value $\overline{R}_{ir} = \pm 12'$ for $z = 80°$ and $\lambda = 4$ m.

Figure 5 shows that for the first model the irregular refraction near the zenith ($z = 0-30°$) is 15-20% of the refraction at the horizon ($z = 80°$), while for the second model the amplitude of the oscillations of refraction at the zenith is close to zero.

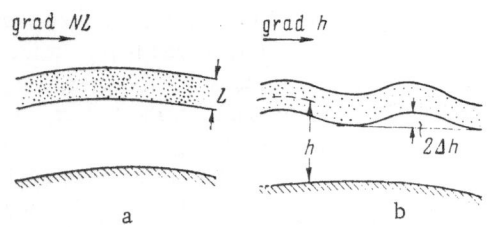

Fig. 4. Two possible models of ionospheric inhomogeneities.

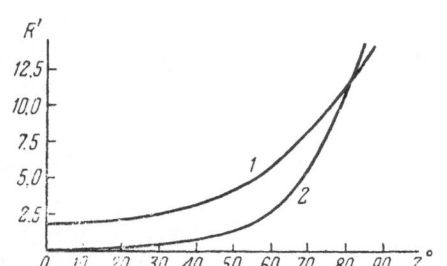

Fig. 5. Dependence of irregular radio wave refraction at λ = 4 m in the ionosphere on zenith angle.

Thus, as the first step in investigation of a model of ionospheric inhomogeneities we can propose measurements of irregular refraction simultaneously at the horizon and near the zenith, where the very presence or absence of oscillations makes it possible to draw definite conclusions.

Investigation of a Model of Ionospheric Inhomogeneities by Simultaneous Measurement of Irregular Vertical Refraction at the Horizon

In 1955-1959 we made a cycle of simultaneous measurements of irregular refraction at zenith angles z = 0-30° and z ~ 77-90° [7].

Measurements Near Horizon. Solar radio spots were used as the emission sources. Measurements were made at the frequency f = 207 Mc/s (λ = 1.45 m) using a marine interferometer by a method similar to that described above. The interferometer antenna was set up at an elevation H = 286.3 m on Mount Koshka (Crimea). The openness of the sea horizon made measurements possible at zenith angles not less than z ~ 77°, i.e., for about one hour after sunrise or before sunset. A simplified method not requiring a knowledge of the coordinates of the radioemitting spots was used in constructing curves of the dependence of refraction on zenith angle on the basis of interference records.

A total of 106 records of sunrise and 61 records of sunset were obtained during the observation period (September 8, 1958-April 20, 1959). The sunset curves differ from the sunrise curves by an almost total absence of refraction oscillations. Only the sunrise curves, examples of which are shown in Fig. 6, were subjected to further detailed analysis. These curves were used in determining the principal characteristics of ionospheric inhomogeneities on the assumption that their height above the earth is h = 300 km. The linear dimensions of the inhomogeneities vary in the range 100-500 km and have a probable value $\overline{d}$ ~ 200-220 km. The amplitudes of the refraction oscillations lie in the range 0.5-5'.0 with the most probable value $\overline{R}_{ir}$ = 2'.5-3'.0.

Measurements Near Zenith. Irregular refraction near the zenith was measured with a horizontal interferometer. Since the method is somewhat different from the preceding method, it will be described in greater detail.

Detection of large ionospheric inhomogeneities by a vertical interferometer is possible only due to angular movement of an emission source. In actuality, the zenith angle covered during one observation period is 10-15° (from z = 90° to z = 30-75°), which corresponds to a linear distance of 800-1200 km along the layer. This means that in the case of a mean linear dimension of inhomogeneities of ~200 km, four to six refraction periods are recorded during one observation period. Thus, the vertical interferometer makes possible the recording of motionless ionospheric inhomogeneities.

In contrast, in observation of extraterrestrial sources near the zenith using a horizontal interferometer, the linear dimensions of the ionospheric region cut off by the aperture of the antenna directional diagrams (usually not more than 15-20°) do not exceed 100-130 km, i.e., less than the mean scale of inhomogeneities of the layer. The detection of large-scale ionospheric inhomogeneities using a horizontal interferometer therefore is possible only if the inhomogeneities move with a sufficiently great velocity.

It is known that there are drifts in the ionosphere whose velocities, measured in numerous studies by observation of small ionospheric inhomogeneities, average ~100 m/sec. Information on movement of large

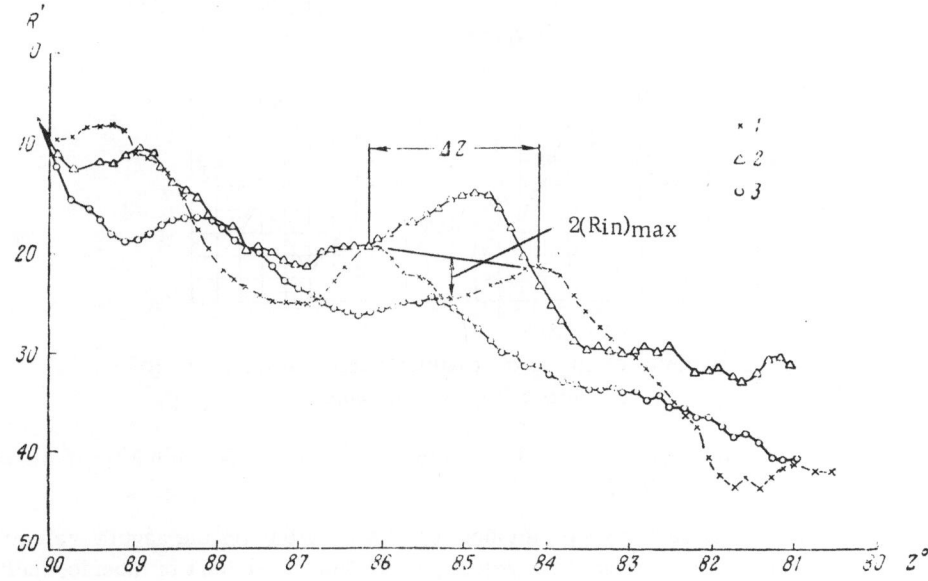

Fig. 6. Samples of curves of irregular refraction $\lambda = 1.45$ m for sunrise hours (z = 77-90°). 1) December 6; 2) December 7; 3) December 8 (1958).

Table 1

Source	Intensity at $f = 52$ Mc/s $\cdot 10^{-23}$ W$\cdot$m^{-2}, cps^{-1}	Coordinates		Observation period	Total No. obs. periods	No. of analyzed records
		declin. δ	right ascension α			
Cassiopeia A	32	58°32′	$23^h 21^m 12^s$	27.IV 1959 1.VI 1959	25	17
Cygnus A	20	40 36	19 58 13	19.III 1959 1.VI 1959	65	50
Taurus A	1.85	22 01	05 31 30	12.XII 1958 1.VI 1959	151	60
Virgo A	2.5	12 41	12 58 06	16.XII 1958 1.VI 1959	145	67

inhomogeneities is more limited. According to [8-10], large-scale inhomogeneities, observed by the pulsed sounding method from the earth, move at velocities of 80-160 m/sec. However, it is not entirely evident that they can be identified with the inhomogeneities responsible for oscillation of refraction. The velocities of movement of large inhomogeneities by the radioastronomical method have been determined in only one study [4]; they were 150-300 m/sec.

In order to detect ionospheric inhomogeneities from their movement, it is necessary to detect at least two or three periods of oscillations of refraction. In the case of the velocities of inhomogeneities mentioned above, the required observation time is 1.0-1.5 h. This means that for the minimum admissible duration of an observation period (1 h) the width of the diagram of each of the interferometer antennas must not be less than 15°. Two parabolic antennas, spaced in an approximately east—west direction at the distance D = 520 m were used in our measurements. The antenna parameters were: eastern — with a right-angle aperture 18 × 22 m; western — with a circular aperture 30 m in diameter. The envelope of the interference diagram is determined by the western antenna. The aperture of its diagram is defined by the relation $\theta = 1.22 \lambda/d$, where d is antenna diameter. Assuming $\theta \approx 15°$, we obtain a minimum wavelength $\lambda = 6$ m. Since elongation of the wave results in a decrease of accuracy of construction of refraction curves (increase of the period of the interference diagram), the working wavelength selected was $\lambda = 5.8$ m.

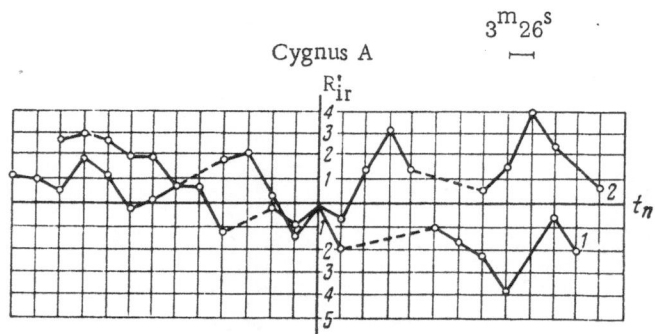

Fig. 7. Samples of curves of irregular refraction for z = 0-30°.
1) April 4; 2) April 6, 1959.

Observations were made from December 12, 1958 to June 1, 1959, in the Crimea (latitude 44°N) using the following cosmic sources. (See Table 1.)

We noted above that in measurements with the horizontal interferometer the angular velocities of movement of the sources were considerably less than the velocity of angular movement of inhomogeneities due to drift. The source therefore can be assumed motionless, as an idealization, and irregular refraction can be considered a function of time, rather than as a function of zenith angle.

We determined the dependence $R_{ir} = f(t)$ for each analyzed observation period. Figure 7 illustrates typical curves. Times of culminations of the sources (indicated for each curve) were used as the zero points on the time scale. The values R_{ir} along the y axis are given in minutes of angle. An irregular shape is characteristic of all the curves.

There are two types of irregularities: short-period and long-period irregularities. Irregularities of the first type are present on virtually all the curves and are detected as random deviations of the values R_{ir} relative to adjacent values. The period of these deviations therefore is less than or equal to $\overline{t_{n+1}} - \overline{t_n}$, where n is the number of the diagram lobe. Their amplitude, averaged for all the records, is approximately ±1'.1. These deviations are unrelated to large-scale ionospheric inhomogeneities and an analysis revealed that they can be caused by the following factors: subjective errors in determining t_n and $\overline{t_n}$ (approximately ±30"), distortions of the interference curve by insignificant industrial and atmospheric interference (approximately ±30") and small-scale ionospheric inhomogeneities (in a period of strong ionospheric disturbance, the angular deviations can attain several minutes of angle [11]).

When there is irregular refraction, several (from 2 to 5) periods of oscillations usually are observed in one observation period; this makes it possible to determine their mean duration and amplitude during the observation period. The duration or period of oscillations varies in the range 5-37 min and has a probable value of 15 min. The dimensions of large ionospheric inhomogeneities cannot be computed directly from this value because this requires data on their velocities, measured simultaneously with the periods.

We will use the results in [8-10] (v = 80-160 m/sec) for estimating the dimensions. Then the dimensions of the inhomogeneities in the direction of their movement is $\overline{d_v} \approx 70-140$ km. If the results in [4] (v = 150-300 m/sec) are used, $\overline{d_v} \approx 135-270$ km. It therefore can be concluded that the dimensions of the observed inhomogeneities are 100-250 km.

The amplitude of the oscillations was determined the same as in measurement of vertical refraction. The most probable value of the amplitude of oscillations $\overline{R}_{ir}$ is ~1'.0, with a corresponding narrow distribution around this value.

Comparison of Measurements of Irregular Refraction Near the Zenith and at the Horizon

The following circumstances must be taken into account when comparing the results of measurements of irregular refraction near the zenith and at the horizon. Due to possible anisometry of ionospheric inhomogeneities in a horizontal direction (see [12]) it is possible to compare only their parameters (scale, etc.), measured in the same direction relative to the directions of the compass. A horizontal interferometer with a base in an east—west direction and a vertical interferometer, oriented at sunrise or sunset, make it possible to measure oscillations with the same angular coordinates of a source, corresponding to a direction along the ionosphere approximately along an east—west line. In this respect, the results of our measurements near the zenith and at the horizon are comparable.

The fact that the ionospheric regions observed using our two interferometers were different and spaced in the indicated direction along the layer at a distance of 1500-2000 km causes some distrust. This circumstance is unimportant in a statistical comparison of the results, but it should be taken into account in a detailed comparison of the individual observation periods.

As pointed out in the prior section, the value of irregular refraction when $\bar{z} = 85°$ for $\lambda = 1.45$ m is $\bar{R}_{ir} =$ 2'.5-3'.0. This value applies to the morning hours. The corresponding value $\bar{R}_{ir}$ for $\lambda = 5.8$ m when z = 85°, computed using the squared dependence of irregular refraction on wavelength ($R_{ir} \approx \lambda^2$) is $\bar{R}_{ir} = 40$-48'. Then, using the curves in Fig. 5, we obtain for $\lambda = 5.8$ m when z = 0-30°: for model 1, $\bar{R}_{ir} = 6'.7$-8'.0, and for model 2, $\bar{R}_{ir} \approx 0$. For comparison of these predicted values and the experimental value of irregular refraction for z = 0-30° at $\lambda = 5.8$ m, the latter also was determined for the morning hours. It was $\bar{R}_{ir} = 0'.9$. This value differs sharply from the predicted value, computed using model 1, and is close to the predicted value for model 2.

A similar comparison of the values of irregular refraction at the horizon and at the zenith for individual observation periods revealed that the relation of values of irregular refraction in the overwhelming majority of cases remains the same as for their probable values. In some cases, the presence of irregular refraction at the horizon was accompanied by its absence at the zenith. On the basis of these facts, it is natural to conclude that large-scale ionospheric inhomogeneities have an undulating structure (model 2) and that there are no significant gradients of the total number of electrons along the layer.

Now we will estimate the parameters of the ionospheric waves, assuming the linear dimension of the inhomogeneities to be $\bar{d} \approx 200$ km, their effective thickness to be $\mathcal{L} \approx 50$ km, and the electron density to be $N = 1.8 \cdot 10^6$ cm^{-3}. The wave amplitude $\Delta\bar{h}$ is determined by the relation which we obtained in [6]:

$$\overline{\Delta h} = \frac{\bar{R}_{ir}\bar{d}^2 \left[1 - \left(\frac{r_0}{r_0 + h}\sin z\right)^2\right]^{3/2}}{L(2\pi)^2 \frac{r_0}{r_0 + h}\sin z}.$$

(2)

Here,

$$L = \mathcal{L}\frac{Ne^2}{2\pi mf^2}.$$

Now we will cite some facts concerning $\mathcal{L}$. We assumed an effective thickness of inhomogeneities $\mathcal{L}$ = 50 km and an effective electron concentration $N = 1.8 \cdot 10^6$ cm^{-3}. With these data, the total number of electrons in a vertical column of the inhomogeneity with a cross section of 1 cm^2 is $\int Ndh = N\mathcal{L} = 9 \cdot 10^{12}$. The corresponding value is $\Delta\bar{h} \approx 0.5$ km. This value agrees approximately with the data in other studies in which the periodic changes of the height of the level of signal reflection (measured by sounding from the earth's surface) are related to large ionospheric inhomogeneities: according to Munro and Bramley [9, 10], $\Delta\bar{h} \approx 0.25$-5.0 km; according to Gusev, Drachev, et al. [8, 13], $\Delta\bar{h}$, computed from the angles of slope of the reflection level (1-2°) and the horizontal scale of inhomogeneities, is 1-2 km. Thus, the value which we obtained may be somewhat too low and therefore we must assume $\Delta\bar{h} \geq 0.5$ km. According to the relation cited above (2), $\Delta\bar{h} \sim 1/L$, where L is the difference between the geometrical and optical thickness of the inhomogeneity. Since the value $N\mathcal{L}$ is related to $\Delta\bar{h}$ by the inverse dependence $N\mathcal{L} \sim 1/\Delta\bar{h}$, taking into account the estimate for $\Delta\bar{h}$ cited above, it must be assumed that $N\mathcal{L} \leq 9 \cdot 10^{12}$. According to the results of rocket investigations, $N\mathcal{L}$ for the entire ionosphere is greater than or equal to $5 \cdot 10^{13}$; according to Gringauz [14], $N \approx 1.8 \cdot 10^6$ cm^{-3} and $\mathcal{L} \geq 350$ km; according to Al'pert et al. [15], $N \approx 1.8 \cdot 10^6$ cm^{-3} and $\mathcal{L} \approx 400$ km.

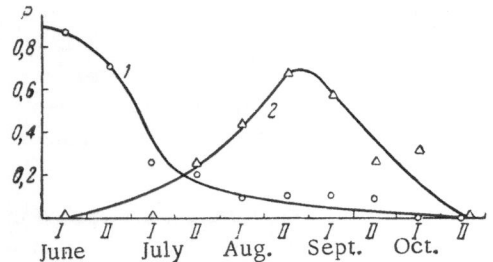

Fig. 8. Frequency of occurrence of refraction irregularities in different months of the year. 1) Using source Taurus; 2) using source Virgo.

Table 2

Source	June (1st half)	Aug. (2nd half)— Sept. (1st half)
Taurus	7^h (max)	1^h
Virgo	14^h	8^h (max)
Sun	5^h	6^h 30^m

The cited data show that L for the inhomogeneities is five or more times lower than for the entire layer. It, therefore, must be assumed that the
densest part of the layer near the ionization maximum
is undulating, since wavelike inhomogeneities are observed by sounding from the earth's surface. In this
case, the effective thickness of inhomogeneities $\mathscr{L}$
= 50 km, which we selected can be somewhat too
great, but apparently it is close to the true value.

Temporal Variations of Irregular
Refraction

The dependence of irregular refraction on time
can be obtained only for the first series of measurements of 1951, because these were made using discrete
sources. For this purpose, all observations of this
series were divided into 15-day groups.

The ratio of the number of days on which refraction irregularities were observed to the total number of observations was determined for each group.
The dependence of this ratio on time is shown in Fig.
8. It is clear that such a clearly expressed temporal
variation is related for the most part to diurnal rather
than to seasonal changes because there is a sharp difference in the curves constructed using data for different
radio stars.

Table 2 gives the approximate local time of observation of sources in the constellations Taurus and Virgo
and the mean time of sunrise for the periods corresponding to the peaks of the curves in Fig. 8.

This table shows that the maximum frequency of occurrence of inhomogeneities is observed 1.5-2 h after
sunrise and, therefore, the mechanism of irregular refraction is related to solar activity.

A similar dependence cannot be obtained using the observations of 1958-1959 because the observations
were made using the sun at the time of its rising and setting, i.e., approximately at the same time of day. However, the fact of an almost total absence of oscillations of refraction in the west at sunset, noted above, is interesting. This possibly is related to the dependence of the degree of ionospheric inhomogeneity on time of
day.

The results obtained in this study make it possible to draw the following principal conclusions. Large-
scale ionospheric inhomogeneities are undulating formations (model 2) with a mean horizontal scale ("period")
$\bar{d} \approx 200$ km and a mean amplitude $\Delta \bar{h} \geq 0.5$ km. Only a part of the layer, with an insignificant effective thickness, constituting $\leq 20\%$ of its total effective thickness, has such undulating curvature. This part apparently is
situated near the region of maximum electron concentration and has a geometrical thickness ≤ 50 km.

Most large-scale ionospheric inhomogeneities are observed in the morning hours (maximum frequency of
occurrence is 1.5-2 h after sunrise), which indicates that they are related to solar activity.

Literature Cited

1.   W. Ross and E. N. Bramley, Nature, 155:162 (1947).
2.   W. Ross and E. N. Bramley, Nature, 164:355 (1949).
3.   E. N. Bramley, Proc. Roy. Soc., 220:39 (1953).
4.   B. M. Chikhachev, Radiotekhn. i Élektron., 5(9):1, 359 (1960).

5. V. V. Vitkevich and Yu. L. Kokurin, Radiotekhn. i Élektron., 2(7): 826 (1957).
6. Yu. L. Kokurin, Radiotekhn. i Élektron., 4(12):1985 (1959).
7. Yu. L. Kokurin, A. N. Sukhanovskii, and Yu. I. Alekseev, Radiotekhn. i Élektron., 6(5):738 (1961).
8. V. D. Gusev, L. A. Drachev, S. R. Mirkotan, Yu. V. Berezin, M. P. Kiyanovskii, M. V. Vinogradova, and T. A. Gailit, Dokl. Akad. Nauk SSSR, 123:817 (1959).
9. G. N. Munro, Proc. Roy. Soc., 202:208 (1950).
10. E. N. Bramley, Proc. Roy, Soc., 220:39 (1953).
11. V. V. Vitkevich and Yu. L. Kokurin, Radiotekhn. i Élektron., 3(11):1373 (1958).
12. V. D. Gusev, S. R. Mirkotan, L. A. Drachev, Yu. V. Berezin, and M.P. Kiyanovskii, Ionospheric Drifts and Inhomogeneities, Collection of Articles, No. 1, Section I of the IGY Program, Izd. Akad. Nauk SSSR (1959).
13. L. A. Drachev and Yu. V. Berezin, Radiotekhn. i Élektron., 2(10):1234 (1957).
14. K. I. Gringauz, Artificial Earth Satellites, 1:62 (1958).
15. Ya. L. Al'pert, É. F. Chudesenko, and B. S. Shapiro, Preliminary Results of Scientific Investigations with the First Soviet Artificial Earth Satellites and Rockets, Collection of Articles, No. 1, Section XI of the IGY Program, Izd. Akad. Nauk SSSR (1958).

MAPS OF RADIO BRIGHTNESS DISTRIBUTION
ON THE SOLAR DISK AT 8 mm

U. V. Khangil'din

A radio telescope with a 22-m parabolic mirror (RT-22) was used in solar observations at 8 mm during August—November 1959. These observations are the next cycle with the RT-22 after the first observations made by Salomonovich in June, 1959.

An article by Salomonovich [1] gives the first two-dimensional distributions of radio brightness on the solar disk at λ = 8 mm. The high accuracy of finishing of the radio telescope mirror and the accuracies attained in setting on angular coordinates make it possible to compile maps of radio brightness distribution in the millimeter wavelength range. This paper gives the two-dimensional distributions on the solar disk at λ = 8 mm obtained on two days of observations (August 13 and October 31, 1959) of the 23 maps compiled for August—November. *

Measurements indicate that the width of the diagram at the 3-dB level in the H plane (corresponding to the cross section in the azimuthal plane) is 1'.7 $\pm$ 0'.1; the width of the diagram in the E plane is determined by measurements with a lesser accuracy. Direct measurements of the width of the diagram using discrete sources at λ = 3 cm make it possible to assume that at both λ = 3 cm and at λ = 8 mm the diagram is very close to axially symmetrical. Measurements and computations of the RT-22 antenna parameters were considered in the above-mentioned article [1].

The maps published here represent the two-dimensional radio brightness distributions over the disk, smoothed due to the finite width of the diagram, superposed on maps of solar activity from the bulletin Solnechnye Dannye (Solar Data) (published by the Main Astronomical Observatory, Pulkovo). The two-dimensional distributions over the disk were determined in several stages, during which the quiet level of the sun was refined successively and serious errors in compilation were detected. It was assumed that the radio disk at λ = 8 mm is equal to the solar photospheric disk. The lines on the map are lines of equal contrast K_i (%). Contrast K_i was determined as $K_i = T_{Ai} - T_{A0}/T_{A0}$, where T_{Ai} is antenna temperature for the solar disk, T_{A0} is antenna temperature corresponding to some level close to the level of the quiet sun (on the maps, the level $K = 0$). Each of the compiled two-dimensional distributions was based on data for one measurement period. The total interval of uncertainty of the quiet level of the sun $2\Delta K$, which is determined in the course of compilation of the maps, is dependent on specific observation conditions on individual days and has values from 2 to 6%.

The mean square deviation from the mean value of the quiet level, determined by averaging for 23 days, was 7%. This value includes both measurement errors and map compilation errors and possible real changes of the quiet level for the days of observation.

The values of the maximum contrast of local regions of increased radio brightness over spot groups attain 40%. On many maps it is possible to detect low-contrast elongated regions over calcium flocculae with contrast values 3-10%. In individual cases over filaments — prominences — there are "dark" local regions; the

---

*Due to the shortness of the article, it is impossible to publish all the compiled maps. All will be published at a later date.

139

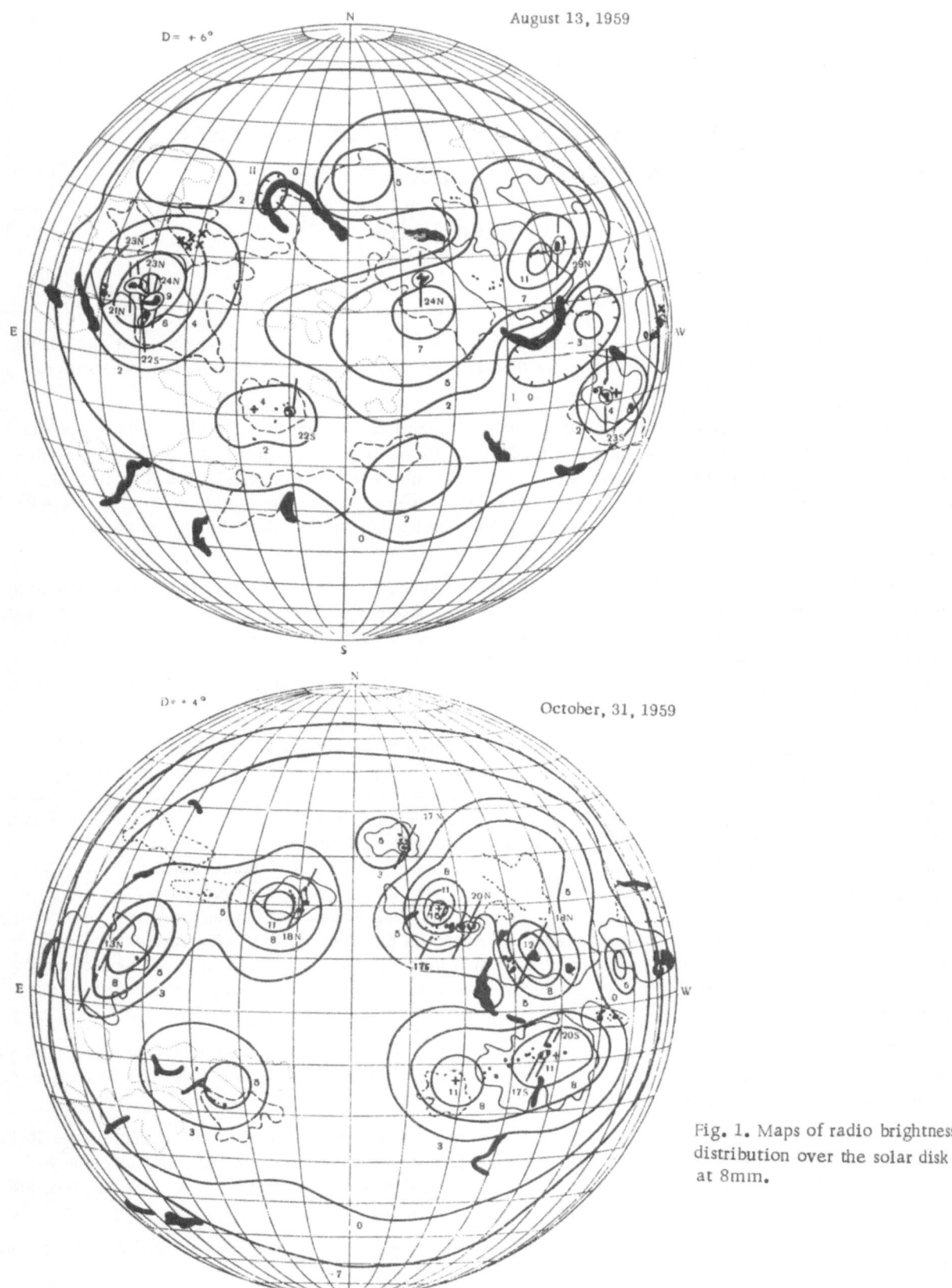

August 13, 1959

October, 31, 1959

Fig. 1. Maps of radio brightness distribution over the solar disk at 8mm.

maximum contrast of such regions on our maps does not exceed 5%. Two such regions can be seen on one of the cited maps. Systematic observations of regions of lesser radio brightness apparently will be possible only with further improvement of apparatus and radio telescopes.

In our earlier article [2], on the basis of the determined two-dimensional distributions at $\lambda = 8$ mm, we gave information on the characteristics of local regions and drew some conclusions concerning the relationship between local regions and active formations on the sun. The accuracy of the compiled maps is higher than for the maps constructed for the first cycle of observations and published in the cited article [1]. This made it possible to draw some new conclusions concerning local regions. These compiled maps, like the maps published earlier, can serve as confirmation of the inherent possibilities of the RT-22 radio telescope for solar investigations in the millimeter range of radio waves.

Literature Cited

1. A. E. Salomonovich, Tr. Fiz. Inst. Akad.Nauk SSSR, 17 (1963).
2. U. V. Khangil'din, Astr. Zh., 41(2): 302 (1964).

ELASTIC DEFORMATIONS OF THE 22–METER
PARABOLIC REFLECTOR OF THE RT-22 RADIO TELESCOPE
OF THE LEBEDEV INSTITUTE (FIAN)
DUE TO GRAVITY LOADING OF THE STRUCTURE

P. D. Kalachev

Introduction

In view of the increasingly more stringent requirements imposed on parabolic reflectors (to lend them greater rigidity), the need is felt to improve the force diagram of the reflector and its mounting on the turntable, as is the need for more exact and more reliable methods in static reflector design in order to determine elastic deformations of the reflector dish, thus improving the design of the load-bearing structures. Bearing in mind the fact that the reflector framework, even if fabricated in a radially symmetric pattern, is a complex interconnected trussed system of girders and bars with a high degree of redundancy, we realize that it is a rather intricate and laborious problem to figure out the elastic deformations of the structure.

For example, the frame of the 22-m radio telescope reflector at the Lebedev Institute (FIAN), the RT-22 [1], built on the simplest radially symmetric plan, is a system which is 37 times statically indeterminate.

A judicious choice of basic structural system calls for a symmetrical design with both symmetric and skew-symmetric loading (for horizontal and vertical positioning, respectively, of a gravity-loaded reflector), reducing the number of redundant unknowns to be determined to 9 or to 5, respectively, and introducing certain simplifying assumptions.

Since the reflector changes its position in space (rotations about the horizontal axis) while in service, appropriate recalculations are needed to secure information on the elastic deformations of the reflector dish at any of its positions. But since the direction of the loading changes in obedience to some law as the reflector rotates about its horizontal axis (this may be either a sinusoidal or cosinusoidal law), the problem is greatly simplified. To be precise, elastic deformations due to the load of the structure dead weight can be determined for any position of the reflector in space by determining the deformations of the two extreme positions of the reflector: the horizontal position when the aperture plane of the reflector dish is horizontal (symmetric loading) and the vertical position when the reflector aperture plane is vertical (skew-symmetric loading). The deformations of the reflector are determined for all other positions as resultants of the component symmetric and skew-symmetric loads so found.

What we are interested in primarily are displacements of the reflecting surface of the reflector on account of elastic deformations in the direction normal to the surface (deflections), since it is precisely these displacements which define the deformed surface of the reflector.

We denote that the displacement of the point K is $\delta_{c(h)}^{k}$ in the horizontal reflector position (Fig. 1), and displacement of the same point and in the same direction of the normal to the surface is $\delta_{kc(v)}^{k}$ with the reflector in the vertical position (Fig. 2).

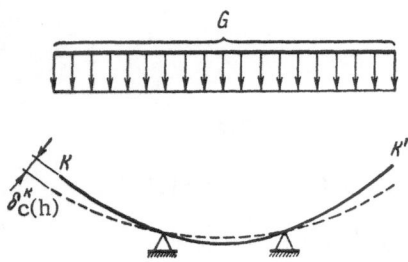

Fig. 1. Diagram of symmetric deformations of reflector dish due to dead weight (gravity loading).

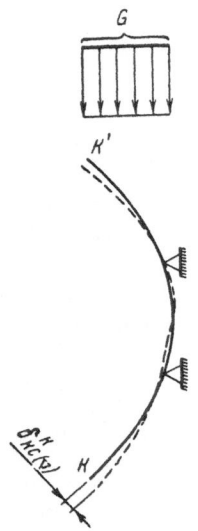

Fig. 2. Diagram of skew-symmetric dead-load deformations of reflector dish.

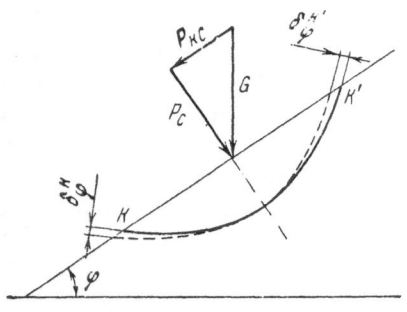

Fig. 3. Diagram of dead-load reflector deformations with reflector in tilted position.

These displacements are proportional to the effective loads:

$$\delta^{k}_{c(h)} = c_1 G \qquad (1)$$

and

$$\delta^{k}_{kc(v)} = c_2 G. \qquad (2)$$

For any intermediate reflector position (Fig. 3), the resultant displacement of point K will be

$$\delta^{K}_{\varphi} = \delta^{K}_{c(h)} \frac{P_c}{G} + \delta^{K}_{Kc(v)} \frac{P_{Kc}}{G} , \qquad (3)$$

but $P_c = G \cos \varphi$ and $P_{kc} = G \sin \varphi$. Substituting the values of P_c and P_{kc} into (3), we get

$$\delta^{K}_{\varphi} = \delta^{K}_{c(h)} \cos \varphi + \delta^{K}_{Kc(v)} \sin \varphi. \qquad (4)$$

For the point K', the resultant displacement $\delta^{k'}_{\varphi}$ will equal the difference

$$\delta^{K'}_{\varphi} = \delta^{K}_{c(h)} \cos \varphi - \delta^{K}_{Kc(v)} \sin \varphi. \qquad (5)$$

The total distortion of the reflecting surface of the dish by elastic deformations is found as the sum of the displacements (deflections) of: the load-bearing elements of the frame, the structural elements (bracing system), and the displacements of the paneling relative to the bracing elements to which the paneling is fastened:

$$\Sigma \delta_{el.d} = \delta_{fr} + \delta_{br} + \delta_p , \qquad (6)$$

where $\Sigma \delta_{el.d}$ is the total displacement of that point of the reflector dish; δ_{fr} is the displacement due to elastic deformations of the frame, δ_{br} is the displacement due to elastic deformations of the bracing structure, and δ_p is the displacement due to elastic deformations of the paneling sheet. As the calculations show, δ_{fr} is usually the basic component with respect to magnitude. The absolute values of the displacements δ_{fr} constitute a function of the frame rigidity and of the magnitude and distribution of the load.

For the reflector considered here, the frame rigidity is determined by the moments of inertia of the load-bearing elements or by their structural height and by the cross sections of the chords and diagonal web members (or by the web cross section in the case of beams).

Structural Design of the Parabolic Reflector and of Its Mount on the Turntable

Parabolic reflectors of radio telescope reflectors and radar stations now in existence are fabricated mostly in a radial symmetry pattern and differ basically in the way the reflector is mounted on the supporting and rotating turntable. The mounting system of the 76-m reflector of the Jodrell Bank radio telescope (Britain) is an un-

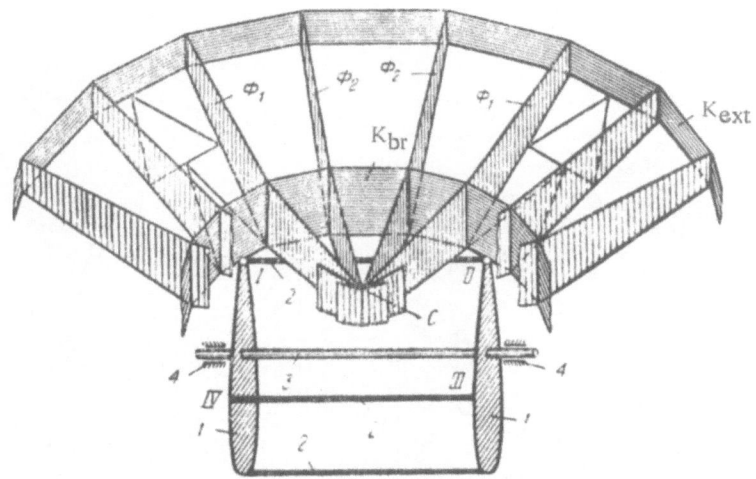

Fig. 4. Design of reflector supporting structure. 1) Rotation sector;
2) rotation sector cross ties; 3) rotation axis about angle of elevation;
4) supporting trunnions. K_{br} is the supporting ring supported at the
four points I, II, III, IV spaced in a radially symmetric pattern with
respect to the ring; C is the central hub of the frame. The sixteen
radial girders supported on the ring and rigidly coupled to the central
hub of the frame are designated as Φ_1 and Φ_2: Φ_1 are the eight girders
placed closer to the reflector supports, and Φ_2 are the eight others
farther removed from the reflector supports.

cantilevered beam supported on both ends (with an elastic support at mid-span) [2]. The 64-m reflector of the
Sydney radio telescope (Australia) is mounted on a developed central hub whose diameter is on the order of
10% of the reflector diameter [3]. This reflector mounting could be viewed as radially symmetric about the
reflector frame.

The 25-m reflector of the Bonn radio telescope (West Germany) is mounted on a supporting ring girder
whose diameter is on the order of 40% of the reflector diameter, and may be likened to a cantilevered beam
supported on two points [4]. The 22-m reflector of the RT-22 radio telescope at FIAN is supported on a ring
girder whose diameter is 50% of the reflector diameter, at four points of support lying in a radially symmetric
pattern about the reflector frame [1].

For a more concrete picture, we consider a way of determining elastic deformations using the 22-m reflector of the RT-22 radio telescope at FIAN as an example.

Design of the Reflector Frame

The mounting of the reflector, supporting ring, and the reflector frame are seen in Figs. 4, 5, and 6.

Horizontal Reflector Position (Symmetric Loading). The reflector frame (Fig. 6) is a
statically indeterminate space truss system with two ring girders (support ring girder and frame ring girder).
Taking the paneling of the ring girder as a diaphragm braced by a single diagonal member, we analyze the
frame from the standpoint of redundancy and geometrical invariability for the general case of loading.

Mid-Portion of Frame Restrained by the Supporting Ring Girder. A single sec-
tored compartment (Fig. 7), which we designate here as No. 1, will be replaced for simplicity in our treatment
by an equivalent compartment (Fig. 8), which is statically determinate and a geometrically rigid space truss.
On adding joint a' to compartment No. 1 (Fig. 9) with the aid of the three bars: a'−b, a'−o, and a'−o', and
then adding joint d' with the aid of the three bars d'−d, d'−a', d'−o', etc., we note that the addition of each

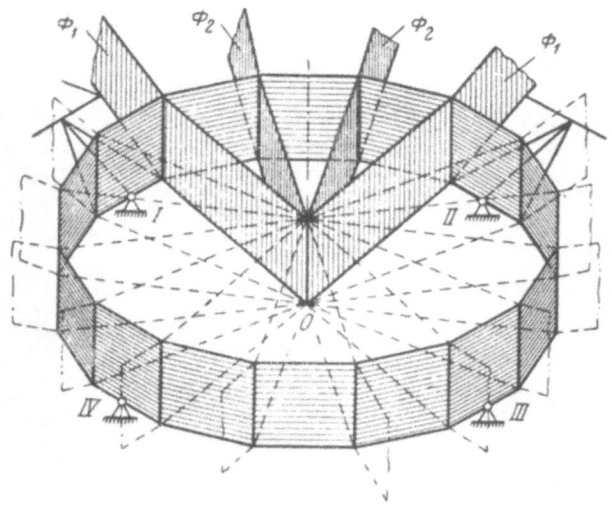

Fig. 5. Design of supporting ring girder. Notation as in Fig. 4.

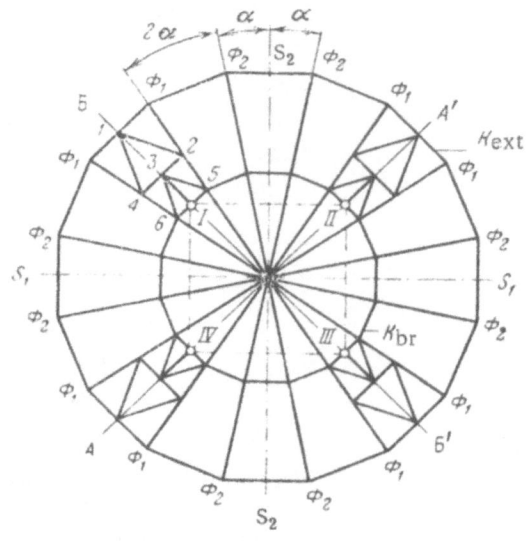

Fig. 6. Design of reflector frame.

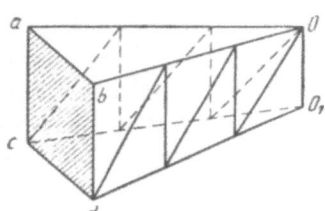

Fig. 7. Compartment of central part of frame.

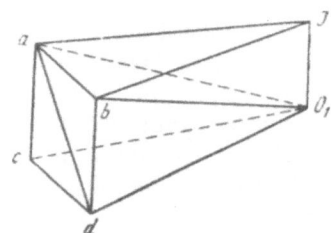

Fig. 8. Equivalent compartment for central part of frame.

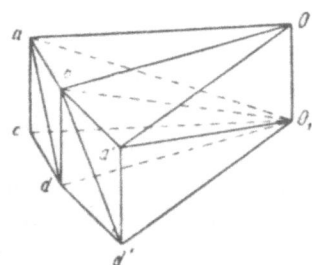

Fig. 9. Composition of central part of reflector frame by successive adjoining of bars.

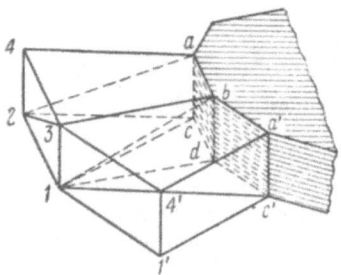

Fig. 10. Composition of peripheral frame of reflector.

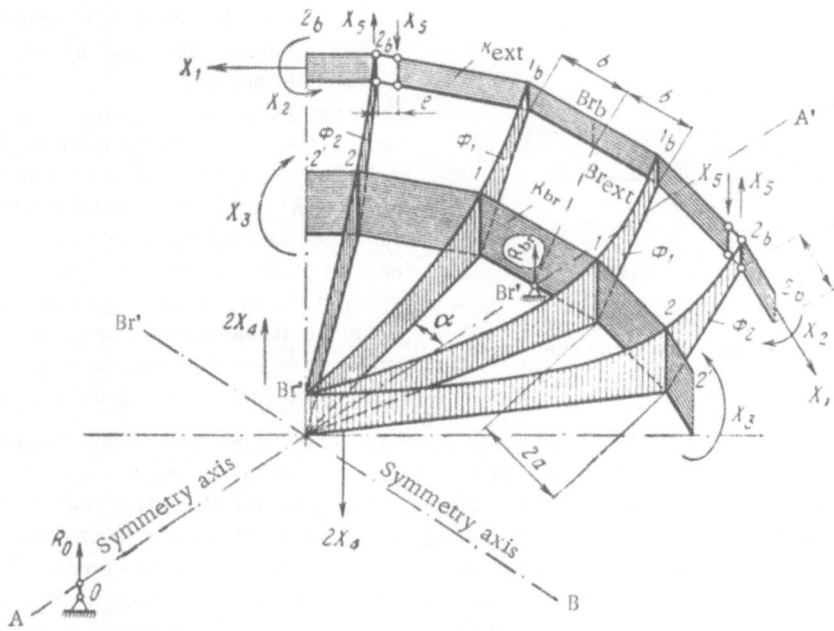

Fig. 11. Part of basic frame system (isometry).

new compartment is accompanied by the introduction of one redundant constraint, the diagonal member b−d'. The formation of the last compartment (No. 16) is accompanied by the introduction of three redundant bars (two horizontal and one diagonal).

The mid-portion of the frame thus contains $Z_{c.r.} = 1 \cdot 14 + 3 = 17$ redundant constraints. Adding the bars 1−b, 1−c, 1−d; 2−a, 2−c, 2−1; 3−b, 3−1, 3−2; 4−2, 4−a, and 4−3 to the mid-portion of the frame, we form a part of the frame with a single diagonal web member (Fig. 10) in the lower panel (1−c).* Further buildup of the frame is accompanied in succession by the addition of one more redundant constraint for each compartment, e.g., joint 4' is formed by the bars 4'−a, 4'−3, and 4'−c', joint 1' is formed by the bars 1'−c', 1−1', and 1'−4', and the redundant bar 4'−1 is added.

The formation of the entire outer portion of the frame, from the supporting ring girder to the outer ring, is accompanied by the addition of redundant members in the amount

$$Z_{ext} = d + \Pi = 15 + 2 + 3 = 20,$$

where d = 15 refers to the diagonal bracing struts (of type 2−3, 1−4', etc.), $\Pi = 2 + 3$ are the closing bars (in the chord of the outer ring girder, of type 3−4, 1−2, cf. Fig. 10). The total number of redundant elements in the frame is

$$Z = Z_{mid} + Z_{ext} = 17 + 20 = 37.$$

It is clear from the analysis that the reflector frame is geometrically rigid and 37 times statically indeterminate.

---

*For simplicity, diagonal member 1−c is introduced in the lower frame panel in place of the lacing in this panel composed of seven bars. The bar 1−c in the plane panel is equivalent to the rods 1−2, 2−3, 3−4, 3−6, 3−5, and 3−1 (cf. Fig. 6) from the standpoint of geometric rigidity and static determinacy.

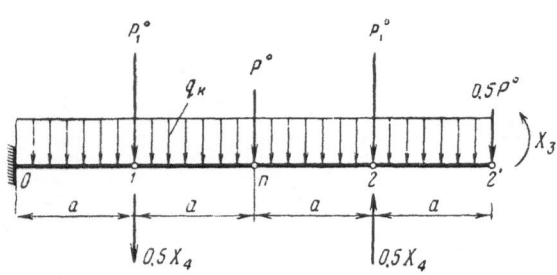

Fig. 12. Basic system for the supporting ring girder.

In horizontal position, the gravity loaded reflector has four symmetry axes (Fig. 6), taking its mounting supports into account: $A-A'$, $B-B'$, S_1-S_1, and S_2-S_2. This means that we need only analyze a part of the frame rather than the entire frame. But since the part of the frame singled out has two symmetry axes of its own: $A-A'$ and $B-B'$, it will be sufficient to analyze one eighth of the frame: $O'-O_b-2_b'-0'$ (Fig. 11), including there the two radial girders Φ_1 and Φ_2, one eighth of the supporting ring girder K_0, and one eighth part of the frame (outer) ring girder K_{ext}, without forgetting, of course, that in general the basic system is one half of the frame resting on two supports (Fig. 11).

Figure 11 shows a part of the basic system in isometric projection: $A-A'$ is the symmetry axis, and R_0 the reaction of the supports. The redundant unknowns are the forces X_1, X_2, X_3, X_4, X_5. The transverse force vanishes in the symmetric cross sections of the outer ring $2'b$, and of the supporting ring $2'$. The normal force also vanishes in the cross sections ($2'$) of the supporting ring, since longitudinal deformations (on the neutral line) of supporting ring beams are zero under symmetric loading. Because of the imperfect symmetry of the reflector frame brought about by the mounting, vertical displacements of the points 1_b and 2_b, as intersections of the outer ring and girders Φ_1 and Φ_2, are unequal, just as in the case of displacements of points 1 and 2, the intersections of the supporting ring with girders Φ_1 and Φ_2. This is what is responsible for the transverse forces X_4 and X_5. The webs of the outer ring girder are cut in the cross sections 2_b and hinge joints are introduced in the chords of the ring girder beam near the intersection of the outer ring girder with the radial girders Φ_2 in such a way that the normal forces X_1 are transmitted through the outer girder (via the chords) and the bending moment X_2 is transmitted through the outer girder, i.e., a virtual four-bar mechanism (with a small value of e) is introduced into the cross section of the outer ring girder.

The bars employed to brace the four symmetrically positioned lower panels (cf. Fig. 6), 1-2, 2-3, 3-4, 3-5, 3-6, 3-1 are inactive under symmetric loading.

Since the rotation angle of the ring girder cross sections is zero (by symmetry) on the ring girder supports, the ring girder may be viewed as a fixed beam on all its supports, and one eighth of the ring in the form of a cantilever fixed at the support O (Fig. 12) may be regarded as the basic system for the supporting ring girder. The loads on the portion of the ring girder sectioned off here are the dead weight of the ring in the form of the uniformly distributed load q [kg/cm],* the load exerted by the radial girders with the adjoining bracing structures and reflector paneling in the form of the concentrated forces P_1^0 [kg],* and the load exerted by the intermediate shortened radial girder, again with the adjoining bracing structure and paneling taken into account, in the form of the concentrated forces $P°$ and $0.5 P°$ [kg].*

Table 1 (see Appendix) gives diagrams of active loads on that portion of the supporting ring girder, bending moment diagrams M [kg · cm], and transverse force diagrams Q [kg]. The bending moment in the supporting ring girder is broken down into two components at the joints where radial girders Φ_1 and Φ_2 intersect the ring, in the direction of succeeding portions of the supporting ring girder and in the direction of the radial girder (Fig. 13). The vectors of the moments acting on portions 2-2', 2-1 of the supporting ring girder and on the radial girder are designated, respectively, as $m_{2-2'}$, m_{2-1}, and m_Φ.

Since the triangle a-2-c is equal to the triangle b-2-c, we have $\overline{a-2} = \overline{b-2}$, i.e.,

$$m_{2-1} = m_{2-2'} = m_K, \tag{7}$$

and

$$m_g = \overline{ac} + \overline{cb} = m_{2-1} \sin \alpha + m_{2-2'} \sin \alpha,$$

---

*[kg] has the dimensionality of force. — Transl. note.

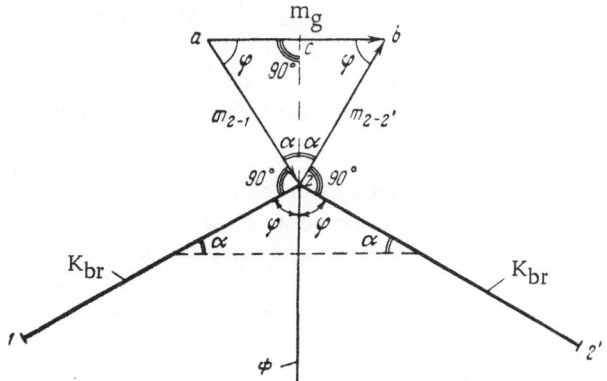

Fig. 13. Diagram showing decomposition of bending moment into components along adjacent portions of supporting ring girder and along radial girder.

or

$$m_g = 2m_\kappa \sin \alpha. \qquad (8)$$

Clearly, from (7), the bending moment in the supporting ring girder does not change its magnitude as we go from one portion (beam) of the supporting ring girder to the adjacent one. This property of the bending moment derives from the symmetry of the structure and applies with equal validity to the outer ring girder. Table 1 therefore considers both ring girders as straight-line beams.

To determine the unknowns X_1, X_2, X_3, X_4, and X_5, i.e., the solutions of the system of canonical equations:

$$\delta_{11}X_1 + \delta_{12}X_2 + \delta_{13}X_3 + \delta_{14}X_4 + \delta_{15}X_5 + \Delta_{1p} = 0;$$
$$\delta_{21}X_1 + \delta_{22}X_2 + \delta_{23}X_3 + \delta_{24}X_4 + \delta_{25}X_5 + \Delta_{2p} = 0;$$
$$\delta_{31}X_1 + \delta_{32}X_2 + \delta_{33}X_3 + \delta_{34}X_4 + \delta_{35}X_5 + \Delta_{3p} = 0;$$
$$\delta_{41}X_1 + \delta_{42}X_2 + \delta_{43}X_3 + \delta_{44}X_4 + \delta_{45}X_5 + \Delta_{4p} = 0;$$
$$\delta_{51}X_1 + \delta_{52}X_2 + \delta_{53}X_3 + \delta_{54}X_4 + \delta_{55}X_5 + \Delta_{5p} = 0, \qquad (9)$$

we have to find the coefficients δ_{ii}, δ_{ik}, and the free terms Δ_{ip}. Using the Maxwell-Mohr method:

$$\delta_{ii} = \int \frac{\overline{M}_i^2}{EI_\kappa} dx + \int \mu \frac{\overline{Q}_i^2}{GF_{\text{web}}} dx + \sum \frac{\overline{N}_i^2 l_i}{EF_i} + \frac{\overline{N}_{\text{ext}}^2 l_{\text{ext}}}{EF_{\text{ext}}};$$

$$\delta_{ik} = \int \frac{\overline{M}_i \overline{M}_k}{EI_k} dx + \int \mu \frac{\overline{Q}_i \overline{Q}_k}{GF_{\text{web}}} dx + \sum \frac{\overline{N}_i \overline{N}_k}{EF_i} l_i;$$

$$\Delta_{ip} = \int \frac{\overline{M}_i M_p}{EI_\kappa} dx + \int \mu \frac{\overline{Q}_i \overline{Q}_p}{GF_{\text{web}}} dx + \sum \frac{\overline{N}_i N_p}{EF_i} l_i, \qquad (10)$$

where the $\overline{M}_i$ and the $\overline{M}_k$ are the bending moments in the beam elements, in this case in both ring girders, on account of unit forces $X_i = 1$ (in the basic system):

M_p is the bending moment in the supporting ring girder due to the gravity loading (dead weight)*; $\overline{Q}_i$ and $\overline{Q}_k$ are the transverse force in the supporting ring girder; Q_p is the transverse force due to the load in the supporting ring girder; E [kg/cm²] is the elastic modulus of the structural material (steel) when stressed normally; G [kg/cm²] is the elastic modulus of the tangentially stressed structural material; I_K [cm⁴] is the moment of inertia of the transverse section through the supporting ring girder and frame (outer) ring girder; F_{web} [cm²] is the cross-sectional area of the web of the supporting ring girder and frame ring girder; μ is the shape factor of the transverse section through the supporting ring girder (μ = 1 in the case of an I-beam); $\overline{N}_i$ and $\overline{N}_k$ are the forces in the radial girder bars due to $X_i = 1$ (in the basic system); N_p [kg] are the forces due to loading in the bars of the radial girders; l_i [cm] are the lengths of the radial girder bars; $\overline{N}_{\text{ext}}$ is the force due to $X_i = 1$ in the chords of the outer ring girder (in the basic system). Terms such as $\int \frac{\overline{M}_i \overline{M}_k}{EI_\kappa} dX$ are displacements due to bending deformations in the two ring girders; $\int \mu \frac{\overline{Q}_i \overline{Q}_k}{GF_{\text{web}}} dX$ are displacements due to transverse

---

* The bending moment and the transverse force due to the dead load of the ring girder are not taken into account in the analysis of that ring.

forces as a result of deformations in the two ring girders; $\sum \dfrac{\overline{N}_i \overline{N}_k l_i}{EF_i}$ are displacements resulting from deformations of the radial girders and of the outer ring girder.

Displacements brought on by deformations of the two ring girders are calculated by the Vereshchagin method, using the appropriate diagrams in Table 1.

To save space, $(\delta_{ik})_k$ denotes displacements due to deformations of the supporting ring girder, $(\delta_{ik})_{ext}$ refers to deformations of the outer frame ring girder, and $(\delta_{ik})_g$ refers to deformations of the radial girders, so that we can write the total displacements (coefficients of the canonical equations) in terms of the polynomials:

$$\delta_{ii} = (\delta_{ii})_k + (\delta_{ii})_{ext} + (\delta_{ii})g;$$
$$\delta_{ik} = (\delta_{ik})_k + (\delta_{ik})_{ext} + (\delta_{ik})g;$$
$$\Delta_{ip} = (\Delta_{ip})_k + (\Delta_{ip})_{ext} + (\Delta_{ip})g. \tag{11}$$

In our case,

$$(\delta_{11})_k = (\delta_{12})_k = (\delta_{13})_k = (\delta_{14})_k = (\delta_{15})_k = (\delta_{22})_k = (\delta_{23})_k = (\delta_{24})_k =$$
$$= (\delta_{25})_k = (\Delta_{1p})_k = (\Delta_{2p})_k = 0,$$

since the forces in the supporting ring girder due to $X_1 = 1$ and $X_2 = 1$ vanish. For the other terms taking deformation of the supporting ring girder into account (14), we have (cf. Table 1):

$$E\,(\delta_{33})_k = \int \frac{\overline{M}_3^2}{I_k}\, dX = \frac{4a^2}{I_k}\;;$$

$$E\,(\delta_{34})_k = \int \frac{\overline{M}_3 \overline{M}_4}{I_k}\, dX = -\frac{2a^2}{I_k}\;;$$

$$E\,(\delta_{35})_k = \int \frac{\overline{M}_3 \overline{M}_5}{I_k}\, dX = \frac{4a^2}{I_k}\;;$$

$$E\,(\delta_{44})_k = \int \frac{\overline{M}_4^2}{I_k}\, dX + \mu k' \int \frac{\overline{Q}_2^4}{F_{web}} dX = \frac{5a^3}{3 I_k} + \mu k' \frac{a}{2 F_{web}};$$

$$E\,(\delta_{45})_k = \int \frac{\overline{M}_4 \overline{M}_5}{I_k}\, dX + \mu k' \int \frac{\overline{Q}_4 \overline{Q}_5}{F_{web}} dX = -\frac{10a^3}{3 I_k} - \mu k' \frac{a}{F_{web}};$$

$$E\,(\delta_{55})_k = \int \frac{\overline{M}_5^2}{I_k}\, dX + \mu k' \int \frac{\overline{Q}_5^2}{F_{web}} dX = \frac{20a^3}{3 I_k} + \mu k' \frac{2a}{F_{web}};$$

$$E\,(\Delta_{3p})_k = \int \frac{\overline{M}_3 M_p}{I_k}\, dX = \frac{-a}{2 I_k}(m_0 + 2m_1 + 2m_n + 2m_2) - \frac{32a^3}{3 I_k} = \frac{-a^2}{I_k}\left(6p^\circ + 5P_1^\circ + \frac{32}{3}qa\right);$$

$$E\,(\Delta_{4p})_k = \int \frac{\overline{M}_4 M_p}{I_k}\, dX + \mu k' \int \frac{\overline{Q}_4 Q_p}{F_{web}} dX + \frac{a^2}{12 I_k}(6m_0 + 11m_1 + 6m_n + m_2)+$$

$$+ \frac{55qa^4}{6 I_k} + \mu k' \frac{a\,(Q_1 + Q_n)}{2 F_{web}} + \mu k' \frac{2qa^2}{F_{web}} = \frac{a^3}{6 I_k}(29P^\circ + 26P_1^\circ + 55qa) + \mu k' \frac{a}{F_{web}}(P^\circ + P_1^\circ + 2qa);$$

$$E\,(\Delta_{5p})_k = \int \frac{\overline{M}_5 M_p}{I_k}\, dX + \mu k' \int \frac{\overline{Q}_5 Q_p}{F_{web}} dX = -\frac{a^3}{3 I_k}(29P^\circ + 26P_1^\circ + 55qa) - \mu k' \frac{2a}{F_{web}}(P^\circ + P_1^\circ + 2qa).$$

For terms taking deformation of the outer ring girder into account, we have

$$(\delta_{12})_{ext} = (\delta_{13})_{ext} = (\delta_{14})_{ext} = (\delta_{15})_{ext} = (\delta_{23})_{ext} = (\delta_{24})_{ext} = (\delta_{33})_{ext} = (\delta_{34})_{ext} =$$
$$= (\delta_{35})_{ext} = (\delta_{44})_{ext} = (\delta_{45})_{ext} = (\Delta_{1p})_{ext} = (\Delta_{2p})_{ext} = (\Delta_{3p})_{ext} = (\Delta_{4p})_{ext} =$$
$$= (\Delta_{5p})_{ext} = 0. \tag{13}$$

For the remaining terms taking deformation of the outer ring girder into account, we have (see Table 1):

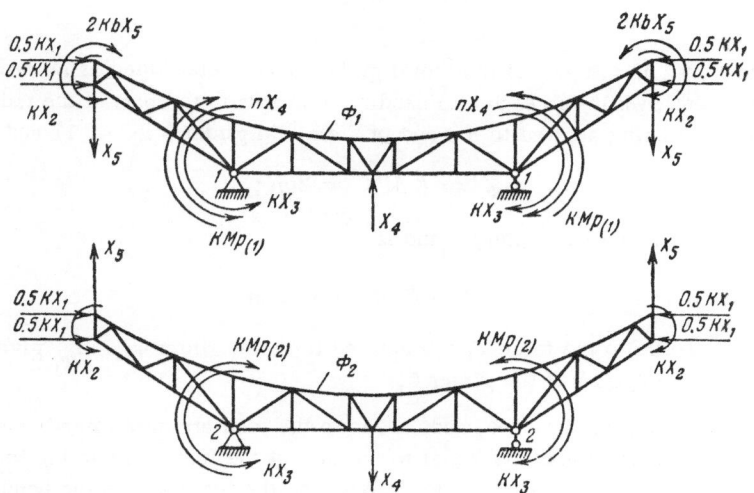

Fig. 14. Diagram of forces acting on the radial girders Φ_1 and Φ_2.

$$E\,(\delta_{11})_{\text{ext}}^{\bullet} = \frac{4b}{F_{\text{ext}}}\,; \quad E\,(\delta_{22})_{\text{ext}} = \frac{4b}{I_{\text{ext}}}\,; \quad E\,(\delta_{25})_{\text{ext}} = \frac{4b^2}{I_{\text{ext}}}\,;$$

$$E\,(\delta_{55})_{\text{ext}} = \frac{20b^3}{3I_{\text{ext}}} + \mu k'\,\frac{2b}{F_{\text{web}}^{\text{ext}}}\,, \tag{14}$$

where k' = E /G, $\mu = 1$, $m_0 = 4a\,(\text{P}^\circ + \text{P}_1^\circ)$; $m_1 = a\,(2.5\text{P}^\circ + 2\text{P}_1^\circ)$; $m_n = a\,(\text{P}^\circ + \text{P}_1^\circ)$; $m_2 = 0.5a\text{P}^\circ$; $Q_{1-n} = \text{P}_1^\circ + 0.5\text{P}^\circ$; $Q_{2-2'} = 0.5\text{P}^\circ$.

We resort to the usual method for determining displacements (in the basic system) due to deformations of the radial girders Φ_1 and Φ_2; the methods are not indicated in the tables owing to lack of space. But the following points are important to bear in mind in view of some features of the system (Fig. 14).

a) Forces in the bars of the radial girders Φ_1 and Φ_2 due to $X_1 = 1$ are respectively identical, i.e., it is sufficient to find the forces due to $X_1 = 1$ in one of the radial girders loaded as in Fig. 16 at the top and bottom end joints by horizontal forces equal to

$$0.5\,KX_1 = 0.5\,K,$$

where K = $2 \sin \alpha$ is the component due to $X_1 = 1$ acting in the plane of the radial girder.

b) Forces in the bars of Φ_1 and Φ_2 due to $X_2 = 1$ are also respectively identical, each radial girder is loaded, on account of $X_2 = 1$, by a moment KX_2 (cf. Fig. 13).

c) Forces due to $X_3 = 1$ appear only in the spans of the radial girders and are also respectively identical in Φ_1 and Φ_2.

d) Forces due to $X_4 = 1$ appear only in bars of the spans of the radial girders, but are different in Φ_1 and in Φ_2; in the case of Φ_2 the forces due to $X_4 = 1$ are determined as in the case of a simple beam supported on two points, while in the case of Φ_1, forces due to the bending moment set up in the supporting ring girder by the force $X_4 = 1$ applied to Φ_2 are added.

The bending moment at point 1 of the supporting ring girder (cf. Fig. 11) due to $X_4 = 1$ is

$$M_1^{\text{к}} = 2a\,0.5 = a. \tag{15}$$

The term $E(\delta_{11})_{\text{ext}}$ corresponds to $N_{\text{ext}}^2\,l_{\text{ext}}/EF_{\text{ext}}$.

The component of this moment acting in the plane of Φ_1 is

$$m_{\Phi_1} = KM_1^{\kappa} = Ka.$$

e) Forces due to $X_5 = 1$ in the bars of the radial girder Φ_2 are determined as in the case of a simple beam supported on two points, while forces due to bending moments in the bars of the radial girder are added — one of the bending moments being applied to the end of the radial girder (cf. Fig. 11 and Fig. 14).

$$m_{\Phi_1}^{\mathrm{ext}} = KM_{1b}^{\mathrm{ext}} = 2bK; \qquad (16)$$

the second bending moment is applied at point 1 and is

$$m_{\Phi_1}^1 = -KM_1^{\kappa} = -2aK. \qquad (17)$$

The bending moment M_1^k is due to the force $X_5 = 1$ applied to radial girder Φ_2. m_{Φ_1}' gives rise to forces only in the bars belonging to the mid-span of the girder Φ_1.

f. Forces in the bars of radial girders Φ_1 and Φ_2 due to the load are determined as follows: in the case of the radial girder Φ_2, forces due to load are equal to the sum of the forces due to the load applied directly to Φ_2 as in the case of a simple beam supported on two points, and the force due to the bending moment set up in the supporting ring girder by the dead load of that ring girder and by a load $0.5\,P^{\circ}$ equal to $m_{\Phi_2}^P = KM_{P(2)}$, where

$$M_{P\,(2)} = 0.5a\,(P^{\circ} + qa); \qquad (18)$$

in the case of radial girder Φ_1, forces due to load are equal to the sum of the forces just as in the case of Φ_2, and to the bending moment set up in the supporting ring girder by the dead load of the girder and by the load on the ring girder, equal (cf. Table 1) to

where $$m_{\Phi_1}^P = KM_{P\,(1)},$$

$$M_{P\,(1)} = m_1 + 4.5qa^2 = a\,(2.5P^{\circ} + 2P_1^{\circ} + 4.5qa). \qquad (19)$$

The displacement due to deformations of the radial girders is given as follows [cf. (13)]:

$$(E\delta_{ii})_g = \Sigma \overline{X}_i^2 \frac{l_i}{F_i}; \quad (E\delta_{ik})_g = \Sigma \overline{X}_i \overline{X}_k \frac{l_i}{F_i}; \quad (E\Delta_{ip})_g = \Sigma \overline{X}_i N_p \frac{l_i}{F_i}, \qquad (20)$$

where $\overline{X}_i$ and $\overline{X}_k$ are the forces in the bars of the radial girders due to unit forces in the basic system; N_p are forces in the same bars due to the load; l_i [cm] and F_i [cm^2] are the lengths and areas of the transverse cross sections of the respective bars in the radial girders.

Comparing respective values in (11), we find

$$E\delta_{11} = \frac{4b}{F_{\mathrm{ext}}} + \Sigma \overline{X}_1^2 \frac{l_i}{F_i};$$

$$E\delta_{12} = \Sigma \overline{X}_1 \overline{X}_2 \frac{l_i}{F_i};$$

$$E\delta_{13} = \Sigma \overline{X}_1 \overline{X}_3 \frac{l_i}{F_i};$$

$$E\delta_{14} = \Sigma \overline{X}_1 \overline{X}_4 \frac{l_i}{F_i};$$

$$E\delta_{15} = \Sigma \overline{X}_1 \overline{X}_5 \frac{l_i}{F_i};$$

$$E\delta_{22} = \frac{4b}{I_{\mathrm{ext}}} + \Sigma \overline{X}_2^2 \frac{l_i}{F_i}; \qquad\qquad E\delta_{23} = \Sigma \overline{X}_2 \overline{X}_3 \frac{l_i}{F_i}; \qquad (21)$$

$$E\delta_{24} = \Sigma \overline{X}_2 \overline{X}_4 \frac{l_i}{F_i} \; ;$$

$$E\delta_{25} = \frac{4b^2}{I_{\text{ext}}} + \Sigma \overline{X}_2 \overline{X}_5 \frac{l_i}{F_i} \; ;$$

$$E\delta_{33} = \frac{4a}{I_k} + \Sigma \overline{X}_3^2 \frac{l_i}{F_i} \; ;$$

$$E\delta_{34} = - \frac{2a^2}{I_k} + \Sigma \overline{X}_3 \overline{X}_4 \frac{l_i}{F_i} \; ;$$

$$E\delta_{35} = \frac{4a^2}{I_k} + \Sigma \overline{X}_3 \overline{X}_5 \frac{l_i}{F_i} \; ;$$

$$E\delta_{44} = \frac{5a^3}{3I_k} + \mu k' \frac{a}{2F_{\text{web}}} + \Sigma \overline{X}_4^2 \frac{l_i}{F_i} \; ;$$

$$E\delta_{45} = - \frac{10a^3}{3I_k} - \mu k' \frac{a}{F_{\text{web}}} + \Sigma \overline{X}_4 \overline{X}_5 \frac{l_i}{F_i} \; ;$$

$$E\delta_{55} = \frac{20a^3}{3I_k} + \mu k' \frac{2a}{F_{\text{web}}} + \frac{20b^3}{3I_{\text{ext}}} + \mu k' \frac{2b}{F_{\text{web}}^{\text{ext}}} + \Delta \overline{X}_5^2 \frac{l_i}{F_i} \; ;$$

$$E\Delta_{1p} = \Sigma \overline{X}_1 N_p \frac{l_i}{F_i} \; ;$$

$$E\Delta_{2p} = \Sigma \overline{X}_2 N_p \frac{l_i}{F_i} \; ; \quad E\Delta_{3p} = - \frac{a^2}{I_{\text{к}}} \left(6p^\circ + 5p_1^\circ + \frac{32}{3} qa \right) + \Sigma \overline{X}_3 N_p \frac{l_i}{F_i} \; ;$$

$$E\Delta_{4p} = \frac{a^3}{6I_{\text{к}}} (29P^\circ + 26P_1^\circ + 55qa) + \mu k' \frac{a}{F_{\text{web}}} (P^\circ + P_1^\circ + 2qa) + \Sigma \overline{X}_4 N_p \frac{l_i}{F_i} \; ;$$

$$E\Delta_{5p} = - \frac{a^3}{3I_{\text{к}}} (29P^\circ + 26P_1^\circ + 55qa) - \mu K' \frac{2a}{F_{\text{web}}} (P^\circ + P_1^\circ + 2qa) + \Sigma \overline{x}_5 N_p \frac{l_i}{F_i} \; . \tag{21}$$

In the case of the reflector dish in question (the 22-m reflector dish of the RT-22 radio telescope of FIAN [Lebedev Institute]), the diameter of the supporting ring girder is $0.5D_{\text{refl}}$, and the a and b are:

$$a = 107.5 \text{ cm}; \quad b = 215 \text{ cm}.$$

The design variables I_k, I_{ext}, F_{ext}, F_{web}, and $F_{\text{web}}^{\text{ext}}$ have the following values: $F_{\text{web}}^{\text{ext}} = 79 \text{ cm}^2$, $I_k = 2.6 \cdot 10^6 \text{ cm}^4$, $I_{\text{ext}} = 1.7 \cdot 10^5 \text{ cm}^4$, $F_{\text{web}} = 184 \text{ cm}^2$, $F_{\text{ext}} = 130 \text{ cm}^2$. The variables P, P°, and q (the load at the joints and the distributed load) are determined by the respective structural dead-load distribution over the joints: $p^\circ = 300 \text{ kg}$, $p_1^\circ = 3020 \text{ kg}$, and $q = 3.5 \text{ kg/cm}$.

Displacements due to unit forces, and loads computed by (21) are as follows:

$E\delta_{11} = 67.494;$	$E\delta_{23} = -0.0003345;$
$E\delta_{12} = 0.16203;$	$E\delta_{24} = 0.0180089;$
$E\delta_{13} = -0.1115;$	$E\delta_{25} = 1.32911;$
$E\delta_{14} = 6.01;$	$E\delta_{33} = 0.000495724;$
$E\delta_{15} = 22.899;$	$E\delta_{34} = -0.02669;$
$E\delta_{22} = 0.006351;$	$E\delta_{35} = -0.018681;$
$E\delta_{44} = 119.208;$	$E\Delta_{2p} = -616.199;$
$E\delta_{45} = 176.3482;$	$E\Delta_{3p} = 280.909;$
$E\delta_{55} = 1306.1;$	$E\Delta_{4p} = -66489.79;$
$E\Delta_{1p} = -227665.5;$	$E\Delta_{5p} = -306535.3.$

$$\tag{22}$$

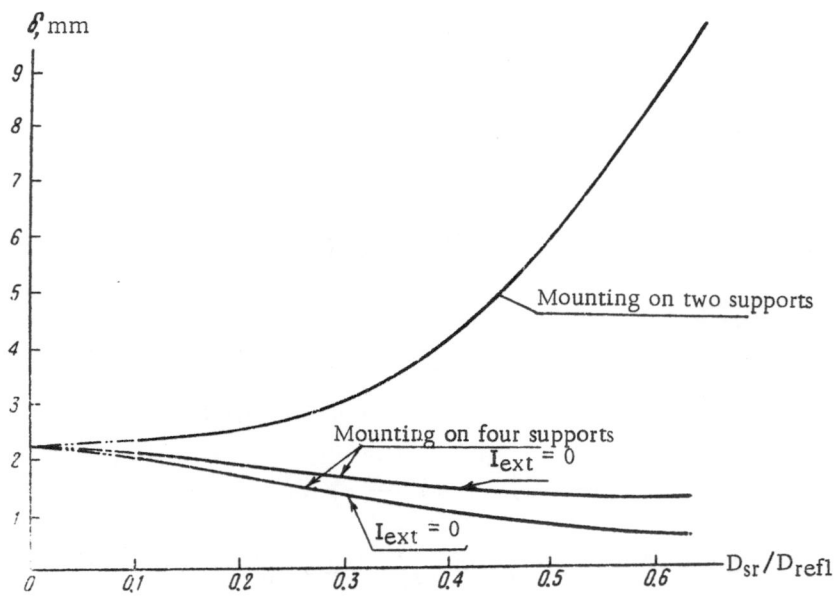

Fig. 15. Diagram of the maximum deflection on the rim of the reflector dish as a function of the relative size of the supporting ring girder. 1) Curve for mounting on two supports; 3) curve for mounting on four supports $(I \neq 0)$; to aid in comparison, we have plotted curve 2 giving the maximum deflections without taking the transverse rigidity of the outer ring girder into account.

In Table 2 of the Appendix we find a matrix useful in solving the system of canonical equations (9), the coefficients attached to the unknowns and free terms of which are taken from (22).

The roots of the equations (redundant unknowns) are:

$$\begin{aligned}
X_1 &= 3846; \\
X_2 &= -20898; \\
X_3 &= 300.796; \\
X_4 &= 186; \\
X_5 &= 168.
\end{aligned} \qquad (23)$$

Vertical displacements of the end points on the cantilevers and of the central point of the radial girders Φ_1 and Φ_2 (deflections) computed in the usual manner* are as follows:

$$\delta_{\Phi_1}^{\kappa} = 0.58 \text{ mm}; \quad \delta_{\Phi_2}^{\kappa} = 0.76 \text{ mm}; \quad \delta_{\Phi_1 \Phi_2}^{\text{trunnion}} = 0.30 \text{ mm}. \qquad (24)$$

Vertical displacements of the supporting ring girder (deflections) at points 1 and 2, similarly computed, are

$$\delta_{1\kappa} = 0.07 \text{ mm}; \quad \delta_{2\kappa} = 0.17 \text{ mm}. \qquad (25)$$

---

*Vertical displacements of the overhang of the radial girder Φ_2 were determined as the sum of the products of the forces in the frame elements determined in the basic system as due to the unit force applied at the point and in the direction of the unknown displacement by the total (computed) forces due to X_1, X_2, X_3, X_4, and X_5, and to the load in the statically indeterminate system.

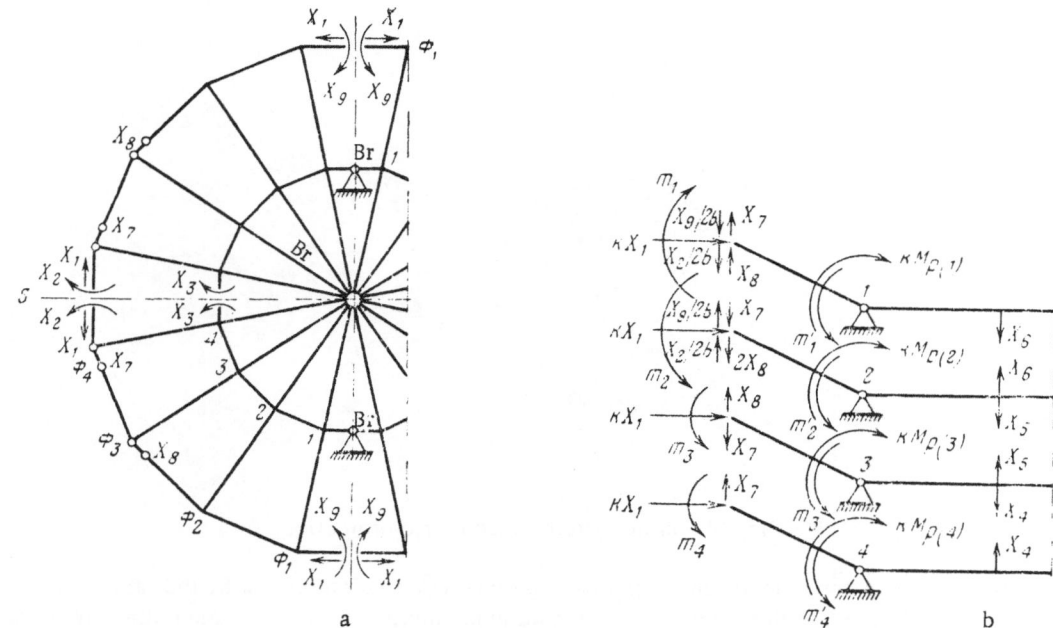

Fig. 16. Basic system for reflector frame mounted on two supports. a) Plan view; b) for radial girders.

Displacements of the reflector frame at the same points were determined in order to evaluate the effect of the transverse rigidity of the outer ring girder, for the case where the moment of inertia of the outer ring girder is negligibly small, i.e., $I_{ext} \approx 0$ (the beam was replaced by its top and bottom chords treated as free bars), and also for the case where the moment of inertia of the outer girder beam was doubled by increasing the structural height (cross-section height) 1.5 times without changing the cross-sectional area (and accordingly without altering the weight of the ring girder). The displacements of the corresponding points are

$$(\delta^K_{\Phi_1})_{I=0} = 0.36 \text{ mm}; \quad (\delta^K_{\Phi_2})_{I=0} = 1.04 \text{ mm}; \quad (\delta_{\Phi_1\Phi_2})_{I=0} = 0.28 \text{ mm};$$

and

$$(\delta^K_{\Phi_1})_{2I} = 0.62 \text{ mm}; \quad (\delta^K_{\Phi_2})_{2I} = 0.74 \text{ mm}; \quad (\delta_{\Phi_1\Phi_2})_{2I} = 0.30 \text{ mm}. \tag{26}$$

It is clear from these results that the presence of the outer ring girder with its optimum transverse rigidity has the effect of substantially adding to the rigidity of the reflector dish under symmetric loading (roughly 30% improvement in rigidity).

Displacements at the above-mentioned points were determined for several values of the ratio D_{sr}/D_{refl} (where D_{sr} is the diameter of the supporting ring girder, D_{refl} is the diameter of the reflector dish) in order to estimate the effect of the size of the supporting ring girder on the rigidity of the reflector, and a diagram (Fig. 15) was plotted of the maximum deflections as a function of the D_{sr}/D_{refl} ratio, i.e.,

$$\delta^K_{\Phi_2} = f\left(\frac{D_{sr}}{D_{refl}}\right).$$

The diagrams clearly show that the maximum symmetric-loading deflection $\delta^K_{\Phi_2}$ for the reflector mounted on four supports varies inversely with the D_{sr}/D_{refl} ratio. As D_{sr}/D_{refl} increases from 0.091 to 0.50, the value of $\delta^K_{\Phi_2}$ drops to roughly one half or one third. Any further increase in the D_{sr}/D_{refl} ratio will have little effect on the maximum deflection $\delta^K_{\Phi_2}$. The diagram shows that the maximum deflections in the case of a two-support mounting are almost seven times the maximum deflections in four-point mounting of the reflector frame.

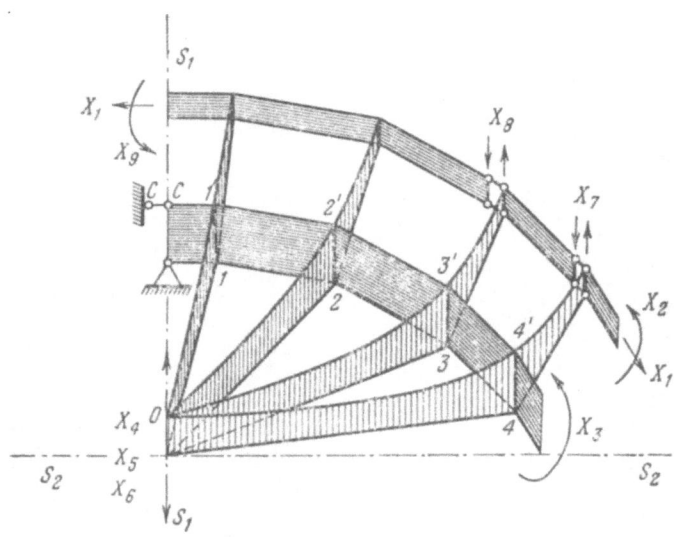

Fig. 17. Basic system in isometric projection.

The displacements for reflector mounting on two supports were determined by the same method as in the four-support case. In this case, the calculations became much more complicated, since the number of redundant unknowns jumped from 5 to 9, i.e., almost doubled.

The basic system for a reflector frame on a two-support mounting including all the load-bearing elements of the frame, except for the bars of the semidiagonals placed in the four bottom panels of the frame (cf. Fig. 6) which are inactive in symmetric loading, is shown in Fig. 16a and Fig. 16b. Taking symmetry into account, the (symmetrically loaded) reflector frame is nine times redundant. Our redundant unknowns are: the normal forces X_1 and the bending moments X_2 and X_9 in the transverse cross sections of the outer ring girder; the bending moments X_3 in the transverse cross sections of the supporting ring girder; the transverse forces X_4, X_5, X_6 constituting the interaction between the radial girders at the center of the frame, and the transverse forces X_7 and X_8 at the points where the web is cut and hinge joints are introduced in the chords of the outer ring beam. The following notation is introduced for forces on the radial girders due to the redundant unknowns and to the load (dead weight).

$$k = 2 \sin \alpha;$$

$$m_1' = ak\left(-\frac{X_2}{b} + \frac{X_3}{a} + X_4 + X_5 + X_6 + 2X_7 + 2X_8 - \frac{X_9}{b}\right);$$

$$m_1 = kX_9;$$

$$m_2 = bk\left(\frac{X_2}{b} - 2X_7 - 2X_8\right);$$

$$m_2' = ak\left(\frac{X_3}{a} + X_4 + X_5 + 2X_7 + 2X_8\right);$$

$$m_3 = bk\left(\frac{X_2}{b} - 2X_7\right);$$

$$m_3' = ak\left(\frac{X_3}{a} + X_4 + 2X_7\right);$$

$$m_4 = kX_2;$$

$$m_4' = kX_3;$$

$$M_{p(1)} = m_1^q + m_1^p;$$
$$M_{p(2)} = m_2^q + m_2^p;$$
$$M_{p(3)} = m_3^q + m_3^p;$$
$$M_{p(4)} = m_4^q + m_4^p;$$

$m_1^q \ldots m_4^q$ and $m_1^p \ldots m_4^p$ are the bending moments in the supporting ring girder due to dead load and due to the weight of the radial girders.

Table 1 (Appendix)

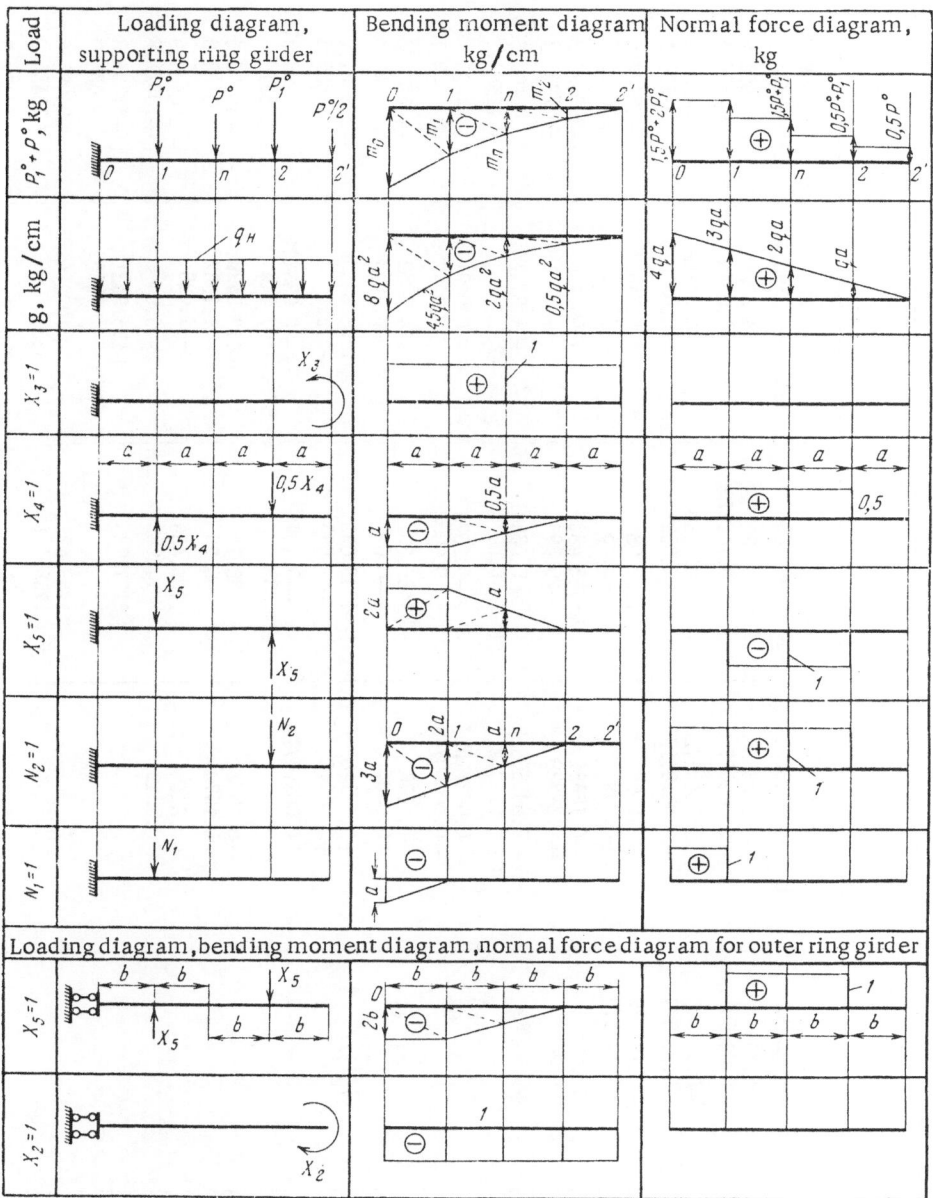

Figure 17 shows the basic system in isometric projection. Only one fourth of the frame is shown, for greater simplicity. At points of support of the frame O–O (cf. Fig. 16a), horizontal bars C–C (Fig. 17) are introduced to restore the effect of the left-hand side discarded in the drawing. The reflector frame exhibits two symmetry axes: S_1-S_1 and S_2-S_2 in accordance with the support mounting.

Since the decrease in the relative size of the diameter of the supporting ring girder all the way to zero (practically to the size of the central hub) renders the two variants of reflector mounting, (four-point and two-point) identical, it is natural that the maximum deflection on the reflector rim will decrease with decrease in the D_{sr}/D_{refl} ratio in the two-support mounting variant and will increase in the four-support variant, as reflected in the diagram.

Table 2

	X_1	$X_2 \cdot 10$	$X_3 \cdot 10$	$X_4 \cdot 10$	X_5	ΔP
I	+67.494	+1.6203 $\alpha=-0.0240065$	−1.115 $\beta=+0.0165199$	+60.1 $\gamma=-0.890449$	+22.899 $\omega=-0.339274$	−227666.5 +3373
II Iα	+0.16203 −0.16203	+0.06351 −0.00389	−0.003345 +0.002677	+0.180089 −0.144279	+1.32911 −0.05497	−616.2 +546.5
III Iβ	−0.1115 +0.1115	−0.003345 +0.002677	+0.00495724 −0.00184197	−0.2669 +0.0993	−0.018681 +0.037829	+280.9 −376.1
IV Iγ	+6.01 −6.01	+0.180089 −0.144279	−0.2669 +0.0993	+1192.08 −5.35	+176.3482 −2.039	−66489.8 +20272.5
V Iω	+22.889 −22.889	+13.2911 −0.5497	−0.18681 +0.37829	+1763.482 −20.39	+1306.1 −7.8	−306535.3 +77241.2
II'	0	+0.05962	−0.000668 $\alpha'=+0.112042$	+0.03581 $\beta'=-6.00637$	+1.27414 $\gamma'=-213.71$	−69.653 +11682.8
III' II'α'	0	−0.000668 +0.000668	+0.00311527 −0.00000748	−0.1676 +0.0004	+0.019148 +0.014276	−95.193 −0.78
IV II'β'	0	+0.03581 −0.03581	−0.1676 +0.0004	+1186.73 −0.02	+174.3092 −0.7653	−46217.3 +41.8
V II'γ'	0	+12.7414 −12.7414	+0.19148 +0.14276	+1743.092 −7.653	+1298.3 −272.3	−229294.1 +14885.5
III''		0	+0.00310779	−0.1672 $\alpha''=+538.0$	+0.033424 $\beta''=-107.549$	−95.973 +308814
IV'' III''α''		0	−0.1672 +0.1672	+1186.71 −8.99	+173.5439 +1.7982	−46175.5 −5163.3
V'' III''β''		0	+0.33424 −0.33424	+1735.439 +17.982	+1.026 −3.6	−214408.6 +10321.8
IV'''			0	+1177.7?	+175.3421 $\alpha'''=-1.48882$	−51338.8 +435.9
V''' IV'''α'''			0	+1753.421 −1753.421	+1022.4 −261	−204086.8 +76434.2
V				0	+761.4	−127652.6

The article on elastic deformations of a parabolic reflector in skew-symmetric loading will show that the optimum size of the supporting ring girder in this case is close to the optimum size of the symmetrically loaded ring girder.

The author expresses his sincere thanks to staff member E. S. Gel'fgat of Mosproekt No. 2 for his valuable comments on the contents of the paper, and to the engineers and designers of the radio astronomy drafting room at FIAN, V. L. Shubeko, V. T. Evdokimova, V. P. Nazarov, and I. A. Emel'yanov, who took part in the calculations and contributed to the shaping of the article.

Literature Cited

1. P. D. Kalachev and A. E. Salomonovich, Radiotekhn. i Élektron., 4(3) (1961).
2. I. Feld, Ann. NY.Acad. Sci., 70(2):155-276.
3. E. G. Bowen and H. C. Minnett, Proc. IRE Australia, 24(2) (1963).
4. L. Mohr, Stahlbau, 27(3) (1958).

ELASTIC DEFORMATIONS DUE TO STRUCTURAL WEIGHT
OF A PARABOLIC REFLECTOR
WITH FOUR-POINT SUSPENSION WHEN IN THE
VERTICAL POSITION (ANTISYMMETRIC LOADING)

P. D. Kalachev

Irrespective of how the reflector is attached to the rotation mechanism, the elastic deformations caused by structural weight are functions of the angle of rotation of the reflector about the horizontal axis, i.e., functions of the attitude, and in general they are a maximum for the vertical position of the reflector (antisymmetric loading).

However, the relation between the magnitudes of the elastic deformations for the symmetric and antisymmetric gravitational loadings, as well as their absolute values, depend very much on the structural details of the attachment of the reflector to the rotation mechanism. The classical layout of the framework of a radially symmetric parabolic reflector will be assumed, and the method of attachment will be sufficient to guarantee the maximum rigidity.

The present article describes a method of calculating the rigidity of the structural framework of a parabolic reflector (antenna) of a radio telescope when in the vertical position and loaded by its own weight (antisymmetric loading). The special feature of the method is that the design of a three-dimensional framework of trusses and beams can be reduced to that for a plane system of trusses and beams.

We will consider a reflector frame consisting of a radially symmetric system fixed at four points (four-point suspension). Figure 1 shows the reflector frame in plan-view; $\Phi_1, \ldots, \Phi_4$ are plane radial trusses; K_H and K_O are the outer and inner ring beams, which consist of plane spacers between each pair of radial trusses $\Phi_1, \ldots, \Phi_4$, etc., and the bracing members P in the lower panels of the frame. The reflector is supported at points I, II, III, IV.

Design Scheme and Basic System

By virtue of the distribution of loads in the elements, a reflector frame subjected to antisymmetric loading can be divided into four parts by two mutually perpendicular planes containing the geometric axis of the reflector. One of these planes is vertical and coincides with the plane YOZ; the other one is horizontal and coincides with the plane XOZ. * Both of these are symmetry planes of the structure. The forces in those structural members that are located symmetrically with respect to the vertical plane are of the same magnitude and sign. Pairs of members located symmetrically with respect to the horizontal plane are subjected to forces of equal magnitude and opposite sign.

---

* The directions of the coordinate axes are clear from Fig. 3. Point O_1, the center of the reflector (in plan), has been taken as origin of coordinates. The axis O_1Z coincides with the geometrical axis of the reflector.

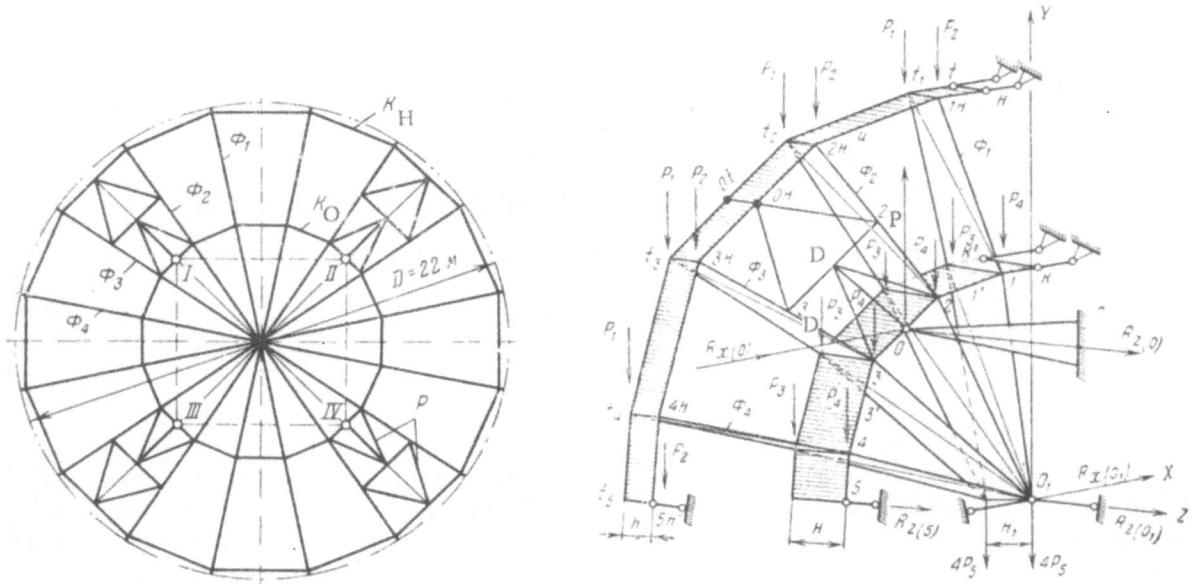

Fig. 1. Layout of reflector framework.   Fig. 2. Equivalent design scheme.

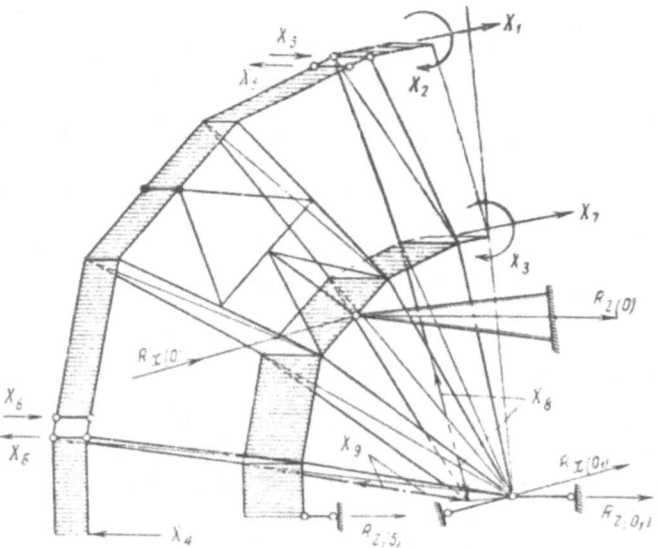

Fig. 3. Basic system.

For this reason, it suffices to design just one quarter of the frame lying in a single quadrant. Assuming a symmetric construction, we will use the design scheme shown in Fig. 2, which is equivalent to that in Fig. 1.

The supporting bars at the points t, H, 5H, K', K, 5, and O_1 in the cut sections through the framework replace the influence of the discarded parts of the framework.

In the general case, the reflector frame is 37-fold statically indeterminate.* By virtue of the symmetric construction, the number of redundant quantities reduces to nine in the case of antisymmetric loading.

Figure 3 depicts the basic system for the reflector framework. The redundant quantities have been chosen to be: the normal force X_1, the bending moment X_2, the shearing force X_4 acting in the cross sections H to t and 5H to t_5 of the outer ring; the shearing forces X_5 and X_6 in the outer ring at the joints with the ends 1H to t_1 and 4H to t_4†; the bending moment X_3 and the normal force X_7 in the supporting section K−K' and the forces X_8 and X_9 in the front chords of the radial frames Φ_1 and Φ_4, respectively. The supports in the basic system are the joints O, O_1, and 5, which are connected to the system through the tie members 3, 2, and 1, respectively, which prevent displacements along the corresponding coordinate axes.

In connection with the decomposition of the spatial truss and girder system into two-dimensional elements, the latter will be assumed to consist of straight beams (for the outer ring in the basic system) of length equal to that between Ht and $5H - t_5$ (see Fig. 2) supported by trusses $\Phi_1, \ldots, \Phi_4$ at the joints 1H, 2H, 3H, and 4H. The fifth support is the joint OH.

The radial trusses receive shearing forces from the outer ring, and joint OH receives longitudinal loads from the outer ring (ring forces).

For the supporting ring the basic system will be a straight girder of length K to 5, which is supported at one of the four main supports O at the reflector, and the auxiliary support 5 introduced in the equivalent design scheme (see Fig. 2).

The loads on the inner ring in the basic system are the loads from the radial trusses $\Phi_1, \ldots, \Phi_4$, obtained by assigning the value unity to the variables X_i, i.e., by setting $X_i = 1$, and from the weight of the radial trusses and the associated constructional elements (grating, boarding, etc.), i.e., from the gravity loads $4 \sum_1^6 P_i$. The loads from the unit values of the variables X_3 and X_7 act directly on the inner ring. The loads from the forces in the bracing of panel $2-3-3H-2H$ (which in the future will be called the "middle panel") arise from the influence of unit values for the variables X_1 and X_7 and from the gravity loads $4 \sum_1^6 P_i$.

In the basic system for the plane radial trusses $\Phi_1, \ldots, \Phi_4$, the supports will be: the inner ring and the central point O_1 of the framework. Shearing forces can be communicated to trusses Φ_1 and Φ_4 only at the central point O_1 and to the trusses Φ_2 and Φ_3 only at the points 2 and 3 on the inner ring (Fig. 2).

a) $X_1 = 1$. We will consider the nature of the influence of the individual forces ($X_1 = 1$) and the loads $\left(4 \sum_1^6 P_i\right)$ on the reflector frame and the forces in the plane element in the basic system $X_1 = 1$.

In view of the structural symmetry, the force $X_1 = 1$ in the outer ring remains constant in the segments $H-1H$, $1H-2H$, and $2H-OH$, and in the segments to the left of joint OH the force vanishes. Moreover, $X_1 = 1$ gives rise to forces in the radial trusses Φ_1, Φ_2, and Φ_3, and in the braces in the middle panel, and in the inner ring.

At the points of intersection between the radial trusses and the outer ring, the component of the ring force $X_1 = 1$, acting in the plane of the radial truss, is $K = 2 \sin \alpha$, where α is half the angle between adjacent radial trusses. In the present case, these components act only in the radial trusses Φ_1 and Φ_2.

---

* See "Elastic Deformations of the 22-Meter Parabolic Reflector of the RT-22 Radio Telescope of the Lebedev Institute (FIAN) due to gravity loading of the Structure," by P. D. Kalachev, this volume, p. 143.

† In the basic system, the wall of the girder in the outer ring has been cut at two places, namely, at the radial trusses Φ_1 and Φ_4, and hinges have been introduced in the flanges of the girder.

The radial trusses Φ_2 and Φ_3 are loaded by transverse forces arising from the bending moment acting in the segment of the outer ring between the trusses Φ_2 and Φ_3. The inner ring is loaded by the respective influences of trusses Φ_1, Φ_2, and Φ_3.

b) $X_2 = 1$. Again, by virtue of the symmetry, the bending moment $X_2 = 1$ remains constant in the segments $\overline{H-1H}$ and $\overline{1H-2H}$ of the outer ring, and at the points of intersection with the trusses Φ_1 and Φ_2 the bending moments in the adjacent segments of the outer ring are balanced by the respective bending-moment reactions acting in the planes of trusses Φ_1 and Φ_2. The radial trusses Φ_2 and Φ_3 are loaded by shearing forces whose reactions balance the moment $X_2 = 1$. The inner ring is loaded in the same way. The forces in the bracing members are zero.

c) $X_3 = 1$. When $X_3 = 1$, the forces in the outer ring, in the brackets of the radial trusses, and in the bracing members of the middle panel, are all zero, except that the inner ring applies loads to the segments of the radial trusses between the rings.

d) $X_4 = 1$. When $X_4 = 1$, the outer ring, the radial trusses Φ_2, Φ_3, and Φ_4, and the inner ring are all loaded. The bracing members in the middle panel remain unstressed.

e) $X_5 = 1$. In this case, the outer ring, the radial trusses Φ_1, Φ_2, and Φ_3, and the inner ring are stressed.

f) $X_6 = 1$. Here, the outer ring, the radial trusses Φ_2, Φ_3, and Φ_4, and the inner ring are loaded.

g) $X_7 = 1$. When $X_7 = 1$, the outer ring (but not the inner ring), the span sections of the radial trusses, and the braces $\overline{2-OD}$, $\overline{3-OD}$, and $\overline{O-OD}$ of the middle panel* are loaded.

h) $X_8 = 1$. The loading $X_8 = 1$ produces stresses in the span sections of the radial trusses Φ_1, Φ_2, and Φ_3, and also in the inner ring.

i) $X_9 = 1$. The loading $X_9 = 1$ produces stresses in the span sections of the radial trusses Φ_2, Φ_3, and Φ_4, and also in the inner ring.

j) $4 \sum_1^6 P_i = 4 (P_1 + \ldots + P_6)$. This loading produces stresses in all members of the framework. The gravity loading of the reflector is applied mainly to the joints of the radial trusses (see Fig. 2). The joint forces $P_1, \ldots, P_4$ can be resolved into components acting in the planes of the radial trusses and in the planes of the segments of the inner and outer rings. We point out here that the cantilevered parts of the radial trusses, $\Phi_1, \ldots, \Phi_4$ and the plane segments $\overline{1H-2H}$ and $\overline{2H-OH}$ of the outer ring are in compression. The segments $OH-3H$ and $3H-4H$ of the outer ring are in tension.

The formulas for the determination of the above forces in the plane members of the framework as a result of $X_i = 1$ and $4 \sum_1^6 P_i$ have been collected in a table.

Table 1 gives the formulas for the determination of the components of the supporting forces along the axes $O_1 X$, $O_1 Y$, and $O_1 Z$ for unit values of the variables X_i and the load $4 \sum_1^6 P_i$. The components of the supporting forces at the supporting point (joint) O_1 in the direction $O_1 Y$ and at the supporting point 5 in the directions of $O_1 X$ and $O_1 Y$ are zero when these points are clamped. The supporting forces from X_5, X_6, X_8, and X_9 also vanish, since each of these forces is self-equilibrating.

Table 2 gives the formulas for the determination of the supporting forces in the outer ring as a result of the $X_i = 1$ and of the loading $4 \sum_1^6 P_i$. The supporting forces caused by X_3, X_7, X_8, and X_9 are zero, since

---

*The braces in the middle panel are $\overline{2-OD}$, $\overline{3-OD}$, $\overline{2D-OH}$, and $\overline{3D-OH}$. In addition, there are the beams $\overline{2-OD}$, $\overline{3-OD}$, and the support tie $\overline{O-OD}$.

Table 1

Supporting joint	R_i	$X_1=1$	$X_2=1$	$X_3=1$	$X_4=1$	$X_7=1$	from loading $4\sum_1^6 P_i$
O	$R_{x(O)}$	$-\left(1+\dfrac{l_K}{l_0}\right)\dfrac{1}{\cos 4\alpha}$	0	0	0	$\dfrac{-1}{\cos 4\alpha}$	$4(P_1+P_2+\ldots P_6)+4\cos 2\alpha\left[(P_1+P_2)\left(1+\dfrac{l_K}{l_0}\right)+P_3+P_4\right]$
	$R_{y(O)}$	0	0	0	0	0	$4(P_1+P_2+\ldots+P_6)$
	$R_{z(O)}$	0	0	0	0	0	$\dfrac{4}{l_0\cos\alpha\cos 4\alpha}[B(P_1+P_2)-P_2 h+P_3 H+P_5 l l_1]$
O_1	$R_{x(O_1)}$	$\left(1+\dfrac{l_K}{l_0}\right)\dfrac{1}{\cos 4\alpha}-1$	0	0	0	$\dfrac{1}{\cos 4\alpha}-1$	$4(P_1+P_2+\ldots+P_6)-4\cos 2\alpha\left[(P_1+P_2)\left(1+\dfrac{l_K}{l_0}\right)+P_3+P_4\right]$
	$R_{y(O_1)}$	0	0	0	0	0	0
	$R_{z(O_1)}$	$\dfrac{0.5h-B}{l_0\cos\alpha}$	$\dfrac{-1}{l_0\cos\alpha}$	$\dfrac{-1}{l_0\cos\alpha}$	$-\dfrac{l_K}{l_0}$	$-\dfrac{H}{2l_0\cos\alpha}$	$\dfrac{-4(1-\cos 4\alpha)}{l_0\cos\alpha\cos 4\alpha}[BP_1+(B-h)P_2+P_3 H+P_5 l I_1]$
5	$R_{z(5)}$	$\dfrac{B-0.5h}{l_0\cos\alpha}$	$\dfrac{1}{l_0\cos\alpha}$	$\dfrac{1}{l_0\cos\alpha}$	$\dfrac{l_K}{l_0}+1$	$\dfrac{H}{2l_0\cos\alpha}$	$\dfrac{-4}{l_0\cos\alpha}[BP_1+(B-h)P_2+P_3 II+P_5 I I_1]$

Table 2

Joint	R_i	$X_1=1$	$X_3=1$	$X_4=1$	$X_5=1$	$X_6=1$	from loading $4\sum_1^6 P_i$
$1H$	R_{1H}	0	0	0	-1	0	0
$2H$	R_{2H}	$-\dfrac{h}{4b}$	$\dfrac{-1}{2b}$	$\dfrac{-1}{2}$	2	1	$-\dfrac{h}{b}P_1(\sin 2\alpha+\cos 2\alpha)$
$3H$	R_{3H}	$\dfrac{h}{4b}$	$\dfrac{1}{2b}$	$\dfrac{1}{2}$	-1	-2	$\dfrac{h}{b}P_1(\sin 2\alpha+\cos 2\alpha)$
$4H$	R_{4H}	0	0	1	0	1	0
OH	R_{OH}	-1	0	0	0	0	$+2(P_1+P_2)(\sin 2\alpha+\cos 2\alpha)$

the outer ring is not affected by these forces. The line of action of all supporting forces of the outer ring (besides R_{OH}) are parallel to the O_1Z axis. The supporting force R_{OH} is parallel to the XOY plane. The signs (+) and (−) refer to positive and negative directions of the components of the supporting forces with respect to the assumed directions of the coordinate axes.

On the basis of what was explained under loading case j), the following formulas are obtained for the determination of the normal forces in the segments of the outer frame:

$$
\begin{aligned}
N_{1H-2H} &= P_2 \tan\alpha; & N_{2H-OH} &= 2P_2 \sin 2\alpha; \\
N_{OH-3H} &= 2P_2 \cos 2\alpha; & N_{3H-4H} &= P_2; \\
N_{t_1-t_2} &= P_1 \tan\alpha; & N_{t_2-_0t} &= 2P_1 \sin 2\alpha; \\
N_{t_3-_0t} &= 2P_1 \cos 2\alpha; & N_{t_3-t_4} &= P_1,
\end{aligned}
\tag{1}
$$

where P_1 and P_2 are the joint forces from the weight of the reflector (see Fig. 2), and 2α is the angle between adjacent radial trusses. The indices attached to N denote the corresponding segments of the outer ring on which the force acts.

The direction of the shearing forces in the radial trusses has been taken in conformity with the line of action of the loading on the given truss.

The supporting forces R_{2H} and R_{3H} of the outer ring (see Table 2) are equal to the corresponding shearing forces on the radial trusses. The signs are, of course, opposite. The forces in the braces Φ_2 and Φ_3 due to the action of $X_1 = 1$ at joint OH are $N_{2-2D}^{(OH)}$ and $N_{3-3D}^{(OH)}$. The components of the resultant forces * in the bracing members $\overline{2-OD}$ and $\overline{3-OD}$ and acting in the planes of Φ_2 and Φ_3 are $N_{2-OD}^{\Phi_2}$ and $N_{3-OD}^{\Phi_3}$. All of these forces are given by the formulas

and

$$
N_{2-2D}^{(OH)} = \frac{-1}{2\cos\gamma\tan\psi}; \qquad N_{3-3D}^{(OH)} = \frac{1}{2\cos\gamma\tan\psi};
$$

$$
\begin{aligned}
N_{2-OD}^{\Phi_2} &= \frac{1+\tan\gamma\cot\psi}{2\cos\gamma\tan\rho} - \frac{l_0+l_k}{2l_0\cos\lambda_n\cos\gamma}; \\
N_{3-OD}^{\Phi_3} &= \frac{1+\tan\gamma\cot\psi}{2\cos\gamma\tan\rho} + \frac{l_0+l_k}{2l_0\cos\lambda_n\cos\gamma}.
\end{aligned}
\tag{2}
$$

The components of the forces in the braces $\overline{2-OD}$ and $\overline{3-OD}$ in the planes of Φ_2 and Φ_3 as a result of $X_7 = 1$ are

$$
(N_{2-OD})_{\Phi_2} = (N_{3-OD})_{\Phi_3} = \frac{\tan 4\alpha}{2\cos\gamma\cos\lambda_n}.
\tag{3}
$$

As a result of loads $4(P_1 + P_2)$ applied to joint OH, the loads in the spars $\overline{2-2D}$ and $\overline{3-3D}$ of frames Φ_2 and Φ_3 are $N_{2-2D}^{(OH)P}$ and $N_{3-3D}^{(OH)P}$; the components of the resultant forces in the braces $\overline{2-OD}$ and $\overline{3-OD}$ are N_{2-OD}^{P} and N_{3-OD}^{P}, which act in the planes of Φ_2 and Φ_3 as a result of both the load $4(P_1 + P_2)$ applied at joint OH as well as the supporting forces are $R_{x(0)}$, $R_{y(0)}$, and $R_{z(0)}$ produced by the load $4\sum_1^6 P_i$. These bracing loads are given by the formulas

$$
\begin{aligned}
N_{2-2D}^{(OH)P} &= (P_1 + P_2)\frac{\sin 2\alpha + \cos 2\alpha}{\cos\gamma\tan\psi}; \\
-N_{3-3D}^{(OH)P} &= (P_1 + P_2)\frac{\sin 2\alpha + \cos 2\alpha}{\cos\gamma\tan\psi};
\end{aligned}
\tag{4}
$$

---

* The loads in the braces $\overline{2-OD}$ and $\overline{3-OD}$ are equal to the sum of the forces in them, i.e., the forces coming from joint OH and the forces coming from joint O_1 and from the supporting forces induced by the force $X_1 = 1$.

$$N_{2-O\,D}^{P} = (N_{2-O\,D}^{R_{x(0)}^{P}} + N_{2-O\,D}^{R_{y(0)}^{P}}) + N_{2-O\,D}^{(OH)\,P};$$

$$N_{3-O\,D}^{P} = (N_{3-O\,D}^{R_{x(0)}^{P}} + N_{3-O\,D}^{R_{y(0)}^{P}}) + N_{3-O\,D}^{(OH)\,P},$$

$$(4)$$

where

$$(N_{2-O\,D}^{R_{x(0)}^{P}} + N_{2-O\,D}^{R_{y(0)}^{P}}) = 2\,\frac{\sin 4\alpha}{\cos \rho \cos \lambda_n}\left\{2\sum_{1}^{6}P_i - \cos 2\alpha\left[(P_1 + P_2)\left(1 + \frac{l_k}{l_0}\right) + P_3 + P_4\right]\right\};$$

$$N_{2-O\,D}^{(OH)\,P} = (P_1 + P_2)\,(\sin 2\alpha + \cos 2\alpha)\,\frac{1 + \tan\gamma\cot\psi}{\sin \rho};$$

$$(N_{3-O\,D}^{R_{x(0)}^{P}} + N_{3-O\,D}^{R_{y(0)}^{P}}) = 2\,\frac{\cos 4\alpha}{\cos \rho \cos \lambda_n}\left\{\cos 2\alpha\left[(P_1 + P_2)\left(1 + \frac{l_k}{l_0}\right) + P_3 + P_4\right] - 2\sum_{1}^{6}P_i\right\}.$$

Table 3 presents formulas for the determination of the supporting forces in the radial trusses $\Phi_1, \ldots, \Phi_4$ as a result of unit values of the variables X_i and as a result of the load $4\sum_{1}^{6}P_i$. The following notation has been adopted in the table:

$$A = B - 0.5h;$$ $$\qquad\qquad B = 1 - \frac{2ak}{l_0};$$

$$\Gamma = \frac{h}{4b}\left(1 + \frac{l_k}{l_0}\right);$$ $$\qquad\qquad K = \frac{k}{l_0};$$

$$И = 3 - \frac{2ak}{l_0};$$ $$\qquad\qquad Л = \frac{hl_k}{4bl_0};$$

$$M = 1 + \frac{ak}{l_0};$$ $$\qquad\qquad \Gamma_1 = 1 + \frac{l_k}{l_0};$$

$$Ж_2 = (N_{2-2\,D}^{(OH)} + N_{2-O\,D}^{\Phi_2})\sin \lambda;$$ $$\qquad\qquad П = \frac{a}{l_0 \cdot \cos \alpha};$$

$$Ж_3 = (N_{3-3\,D}^{(OH)} + N_{3-O\,D}^{\Phi_3})\sin \lambda;$$ $$\qquad\qquad П_1 = \sin \omega_{II} \cos 2\alpha;$$

$$Ш = (N_{2-2\,D}^{(OH)\,P} + N_{2-O\,D})\sin \lambda;$$ $$\qquad\qquad T = N_{t_1}^{\Phi_1} + N_{1H}^{\Phi_1};$$

$$Ш_1 = \frac{B}{l_0}\left[N_{t_2}^{\Phi_2} + \left(1 - \frac{h}{B}\right)N_{2H}^{\Phi_2} + \frac{H}{B}\,N_{D_2}^{\Phi_2} + \frac{H_1}{B}\,P_{DO_2}^{\Phi_2}\right];$$

$$Ш_2 = BP_1 + (B - h)\,P_2 + HP_3;$$

$$Ш_3 = \frac{B}{l_0}\left[N_{t_3}^{\Phi_3} + \left(1 - \frac{h}{B}\right)N_{3H}^{\Phi_3} + \frac{H}{B}\,N_{D_3}^{\Phi_3} - \frac{H_1}{B}\,P_{DO_2}^{\Phi_3}\right];$$

$$Ш_4 = (N_{3-3\,D}^{(OH)\,P} + N_{3-O\,D}^{P\,\Phi_3})\sin \lambda;$$ $$\qquad Ш_5 = \frac{m_4^{\Phi_3} + m_5^{\Phi_3}}{l_0},$$

where

$$(N_{2-2\,D}^{(OH)} + N_{2-O\,D}^{\Phi_2}) = \frac{1}{2\cos\gamma\tan\phi} + \frac{1 + \tan\gamma\cot\psi}{2\cos\gamma\tan\rho} - \frac{l_0 + l_k}{2l_0\cos\gamma\cos\lambda_n}$$

$$(N_{3-3\,D}^{(OH)} + N_{3-O\,D}^{\Phi_3}) = \frac{1}{2\cos\gamma\tan\phi} + \frac{1 + \tan\gamma\cot\psi}{2\cos\gamma\tan\rho} + \frac{l_0 + l_k}{2l_0\cos\gamma\cos\lambda_n};$$

$$N_{2-2\,D}^{(OH)\,P} = (P_1 + P_2)\,\frac{\sin 2\alpha + \cos 2\alpha}{\cos\gamma\tan\psi};$$

$$N_{3-3\,D}^{(OH)\,P} = (P_1 + P_2)\,\frac{\sin 2\alpha + \cos 2\alpha}{\cos\gamma\tan\psi};$$

Table 3

Truss No.	R_i	$X_1 = 1$	$X_2 = 1$	$X_3 = 1$	$X_4 = 1$	$X_5 = 1$
Φ_1	$R_1^к$	$КА$	$К$	$К$	$О$	$Г_1$
	R_1^q	$КА$	$К$	$К$	$О$	$R_1^к = 1$
Φ_2	$R_2^к$	$-КАБ + Г + Ж_2$	$\frac{1}{2b}Г_1 - КБ$	$КБ$	$\frac{1}{2}Г_1$	$2(Г_1M - bК)$
	R_2^q	$R_{2H} - R_2^к + Ж_2$	$\frac{l_к}{2bl_0} - КБ$	$КБ$	$\frac{l_к}{2l_0}$	$R_2^к - 2$
Φ_3	$R_3^к$	$ПКАИ + Г + Ж_3$	$\frac{1}{2b}Г_1 + ПКИ$	$ПКИ$	$\frac{1}{2}Г_1 + КБ(aГ - b)$	$Г_1$
	R_3^q	$R_3^к - R_{3H} - Ж_3$	$\frac{l_к}{2bl_0} + ПКИ$	$ПКИ$	$R_3^к - R_{3H}$	$Г_1$
Φ_4	$R_4^к$	$ПКА$	$ПК$	$ПК$	$Г_1M - bК$	$О$
	R_4^q	$ПКА$	$ПК$	$ПК$	$R_4^к - 1$	$О$

$$N_{2-OD}^P = \left[(P_1 + P_2)(\sin 2\alpha + \cos 2\alpha)\frac{1 + \tan\gamma \cot\psi}{\sin\rho} + \right.$$

$$\left. + \frac{2\sin 4\alpha}{\cos\rho \cos\lambda_n}\left\{2\sum_1^6 P_i - \cos 2\alpha\left[(P_1 + P_2)\left(1 + \frac{l_k}{l_0}\right) + P_3 + P_4\right]\right\}\right]\frac{\cos\rho}{\cos\gamma};$$

$$N_{3-OD}^P = \left[(P_1 + P_2)(\sin 2\alpha + \cos 2\alpha)\frac{1 + \tan\gamma \cot\psi}{\sin\rho} + \right.$$

$$\left. + \frac{2\cos 4\alpha}{\cos\rho \cos\lambda_n}\left\{\cos 2\alpha\left[(P_1 + P_2)\left(1 + \frac{l_k}{l_0}\right) + P_3 + P_4\right] - 2\sum_1^6 P_i\right\}\right]\frac{\cos\rho}{\cos\gamma}.$$

The notation in formulas (2), (3), and (4) is as follows: γ is the angle between bar $\overline{2D-2H}$ and the normal to bar $\overline{2D-3D}$; ψ is the angle between bar $\overline{2D-OH}$ and the same normal; ρ is the angle between bar $\overline{2-OD}$ and the same normal; ω_D is the angle between the rear spars of the inner part $\overline{O_1-2}$ of the radial truss and the front spar of this same part of the truss; λ is the angle between the bars $\overline{2-2H}$ and $\overline{O_1-2}$, and λ_D is the angle between the bar $\overline{O-OD}$ and the XOY plane.

$$N_{t_1}^{\Phi_1} = P_1\frac{\cos 2\alpha}{\cos\alpha}; \qquad N_{1H}^{\Phi_1} = P_2\frac{\cos 2\alpha}{\cos\alpha};$$

$$N_{t_2}^{\Phi_2} = 2P_1\sin\alpha(2\cos 2\alpha - \tan\alpha); \quad N_{2H}^{\Phi_2} = 2P_2\sin\alpha(2\cos 2\alpha - \tan\alpha);$$

$$N_{t_3}^{\Phi_3} = 2P_1\cos 5\alpha; \quad N_{3H}^{\Phi_3} = 2P_2\cos 5\alpha;$$

$$N_{t_4}^{\Phi_4} = 2P_1\sin\alpha; \quad N_{4H}^{\Phi_4} = 2P_2\sin\alpha; \quad N_{D2}^{\Phi_2} = 2P_3\sin\alpha(2\cos 2\alpha - \tan\alpha);$$

$$N_{D3}^{\Phi_3} = 2P_3\sin 3\alpha; \quad P_{DO_2}^{\Phi_2} = 4P_5\frac{\cos 3\alpha}{\cos 6\alpha}; \quad P_{DO_2}^{\Phi_3} = 4P_5\frac{\sin 3\alpha}{\cos 6\alpha};$$

$$m_1^{\Phi_3} = 2akR_1^k; \quad m_4^{\Phi_3} = 2akR_4^k; \quad m_5^{\Phi_3} = 3akR_{z(5)}; \quad k = 2\sin\alpha.$$

Explanation of indices: a) In $N_{t_i}^{\Phi_i}$ and $N_{iH}^{\Phi_i}$, the superscript Φ_i denotes the number of the radial truss $i = 1, \ldots, 4$ in the plane of which the force acts, and the subscript t_i or iH denotes the joint to which the load P is applied (t_i and iH are on the front and rear chords, respectively, of the radial trusses).

$X_6=1$	$X_7=1$	$X_8=1$	$X_9=1$	Load $4\sum_1^6 i$
0	$\frac{1}{2}HK$	$\frac{H_1}{l_0}\cos\omega_{II}$	0	$\frac{B}{l_0}\left(T+\frac{H}{B}N_{D_1}^{\Phi_1}-\frac{h}{B}N_{1H}^{\Phi_1}\right)$
0	$\frac{1}{2}HK$	$R_1^K+\sin\omega_{II}$	0	$\frac{B}{l_0}\left(T+\frac{H}{B}N_{D_1}^{\Phi_1}+\frac{h}{B}N_{1H}^{\Phi}\right)$
Γ_1	$\frac{1}{2}\left(HKБ+\frac{\tan\lambda_{II}}{\cot 4\alpha}\right)$	$2R_1^K$	$\frac{H_1}{l_0}\cos\omega_{II}$	$\Gamma_1 R_{2H}+III+III_1-\frac{m_1^{\Phi_2}}{l_0}$
Γ_1-1	$\frac{1}{2}HKБ$	$R_2^K+2II_1$	$R_2^K+\sin\omega_{II}$	$R_2^K-R_{2H}-III$
$2(\Gamma_1 M-bK)$	$\frac{1}{2}(HПKИ+\tan\lambda_{II})$	$\frac{H_1}{l_0}\cos\omega_{II}$	$2R_4^K$	$\Gamma_1 R_{3H}-III_3+III_4+III_5$
R_3^K-2	$\frac{1}{2}HПKИ$	$R_3^K+\sin\omega_{II}$	$R_3^K+2II_1$	$R_3^K-R_{3H}-III_4$
Γ_1	$\frac{1}{2}HПK$		$\frac{H_1}{l_0}\cos\omega_D$	$K(III_2-aR_{z(5)})$
Γ_1-1	$\frac{1}{2}HПK$	0	$R_4^K+\sin\omega_D$	$K(III_2-aR_{z(5)})$

b) In the case of m, the superscript Φ_k denotes the number of the radial truss on which the bending moment $m_i^{\Phi_k}$ acts and the subscript i denotes the radial truss whose supporting forces produce the bending moment $m_i^{\Phi_k}$. If $i=5$, then $m_5^{\Phi_k}$ is caused by the supporting force at point 5.

c) As for the indices on R, the superscripts k and Q denote the supporting points: k is support on the ring and Q is support in the central point. The suffices i denote the number of the radial truss (for all forms of loading).

For checking the validity of the calculated values of the supporting forces at point Q of the radial trusses (for all forms of loading), we have the relation

$$\sum_1^4 R_i^Q = R_{z(O_1)},$$

(5)

where $R_{z(O_1)}$ is given in Table 1.

The system of canonical equations relating the redundant variables X_i has the familiar form

$$\delta_{11}X_1 + \delta_{12}X_2 + \ldots + \delta_{19}X_9 + \Delta_{1p} = 0;$$
$$\delta_{21}X_1 + \delta_{22}X_2 + \ldots + \delta_{29}X_9 + \Delta_{2p} = 0;$$
$$\cdots\cdots\cdots\cdots\cdots\cdots\cdots\cdots\cdots\cdots\cdots$$
$$\cdots\cdots\cdots\cdots\cdots\cdots\cdots\cdots\cdots\cdots\cdots$$
$$\delta_{91}X_1 + \delta_{92}X_2 + \ldots + \delta_{99}X_9 + \Delta_{9p} = 0,$$

(6)

where the coefficients and the constants, which are the displacements caused by unit values of the forces and loads, can be determined by the Maxwell-Mohr formulas

Table 4

Deflection	Truss				Deflection	Truss			
	Φ_1	Φ_2	Φ_3	Φ_4		Φ_1	Φ_2	Φ_3	Φ_4
Cantilevers δ_i^{κ}, mm. . .	3.30	2.80	1.50	0.42	Inner ring δ_{i1}^{ext}, mm. .	1.55	1.30	0.88	0.31
Inner ring δ_i^{σ}, mm. .	0.28	0.13	—0.03	—0.02	Cantilevers δ_{i2}^{κ}, mm. .	0.00	0.20	—0.26	—0.22
Cantilevers δ_{i1}^{κ}, mm. .	3.30	2.60	1.76	0.62	Inner ring δ_{i2}^{ext}, mm. .	—1.27	—1.17	—0.91	—0.33

$$\delta_{ii} = \left(\Sigma \overline{X}_i^2 \frac{l_i}{EF_i} \right) + \Sigma \int \frac{M_i^2}{EI} dX + \mu \Sigma \int \frac{\overline{Q}_i^2}{GF_{\text{web}}} dX;$$

$$\delta_{ik} = \left(\Sigma \overline{X}_i \overline{X}_k \frac{l_i}{EF_i} \right) + \Sigma \int \frac{\overline{M}_i \overline{M}_k}{EI} dX + \mu \Sigma \int \frac{\overline{Q}_i \overline{Q}_k}{GF_{\text{web}}} dX;$$

$$\Delta_{ip} = \left(\Sigma \overline{X}_i N_p \frac{l_i}{EF_i} \right) + \Sigma \int \frac{\overline{M}_i M_p}{EI} dX + \mu \Sigma \int \frac{\overline{Q}_i Q_p}{GF_{\text{web}}} dX. \tag{7}$$

In these formulas, the first terms on the right-hand sides are the displacements caused by the deformations of the bars in the radial trusses and by the deformations in the outer ring as a result of the normal forces. The second terms are the displacements caused by the bending moments, and the third terms are the displacements as a result of the transverse forces in the outer and inner rings. The symbols $\overline{X}_i$, $\overline{X}_k$, and N_p denote the normal forces in the bars and in the outer ring; $\overline{M}_i$, $\overline{M}_k$, and M_p are the bending moments in the outer and inner rings; $\overline{Q}_i$, $\overline{Q}_k$, and Q_p are the shearing forces in the outer and inner rings as a result of $X_i = 1$ and $4 \sum_1^6 P_i$; l_i, F_i, F_{web}, and I are geometric parameters, and E and G are the elastic moduli of the structural material. In the usual way, the design forces in the elements of the framework are given by

$$N_{\text{des}} = \sum_1^9 \overline{X}_i X_i + N_{P_i}, \tag{8}$$

where X_i are the values of the unknowns found from equations (7), and N_{P_i} are the forces in the respective elements of the framework as a result of the loading. The deflection at an arbitrary point of the framework is determined in the usual way:

$$\delta_n = \Sigma \overline{N}_n N_{\text{des}} \frac{l_i}{EF_i} + \Sigma \int \frac{\overline{M}_n M_{\text{des}}}{EI} dX + \mu \Sigma \int \frac{\overline{Q}_n Q_{\text{des}}}{GF_{\text{cr}}} dX, \tag{9}$$

where $\overline{N}_n$, $\overline{M}_n$, and $\overline{Q}_n$ are the normal forces, the bending moments, and the shearing forces, respectively, that arise in the elements of the framework as a result of a unit force applied at the point where the deflection is required, and N_{des}, M_{des}, and Q_{des} are the corresponding design values of the normal forces, bending moments, and shearing forces in the same elements of the framework. The above-described method was used for the calculation of the rigidity of the 22-m radio-telescope reflector FIAN RT-22.

Table 4 gives the deflections of eight points on the framework: at the end points of the cantilevered radial trusses $\Phi_1, \ldots, \Phi_4$, and at four points on the inner ring (at the points of intersection of the inner ring with the radial trusses). The analysis of the transverse displacements (deflections) of the above points on the reflector frame shows that they can be decomposed into two groups.

1. The displacements δ_{i1}, whose magnitudes are proportional to the distance of the considered point of the framework from the horizontal axis of symmetry of the reflector, are given by

$$\delta_{i1} = K_D \rho_i, \tag{10}$$

where K_D is the coefficient of proportionality, ρ_i is the distance of the point from the above horizontal axis, δ_{i1} are those deflections of the points of the framework as a result of which the shape of the reflector is not distorted but simply rotates as a whole about the horizontal axis. The coefficient of proportionality K_D is equal to the angle of rigid rotation of the reflector (rotation of the antenna aperture).

2. The displacements δ_{i2} equal the differences

$$\delta_{i2} = \delta_i - \delta_{i1}. \tag{11}$$

The displacements δ_{i2} characterize the actual deformation of the reflector due to structural weight in the case of antisymmetric loading (vertical position of the reflector).